Applied ethology 2010: Coping in large groups

Proceedings of the 44th Congress of the International Society for Applied Ethology (ISAE)

Coping in large groups

Swedish University of Agricultural Sciences
Uppsala, Sweden
4-7 August 2010

edited by:
Lena Lidfors
Harry Blokhuis
Linda Keeling

Wageningen Academic
P u b l i s h e r s

ISBN 978-90-8686-150-7

First published, 2010

Welcome

Research in applied ethology has a long tradition at the Swedish University of Agricultural Sciences, as have the links to this society. This is the third time that the international congress has been organised by our department; the previous years were 1978 and 1988. The name of the society has changed, as has the name of our department, but the tradition continues from one 'generation' of scientists to another. It is a privilege and honour to host the congress again this year.

The conference is organised around the main theme of coping in large groups. We feel this is an important topic, not only for researchers working with farm animals, but also for those working with laboratory, companion, sport and other categories of animals. It is also a topic of basic biological interest. The Wood-Gush Memorial Lecture will be presented by Charlotte Hemelrijk, professor at the University of Groningen in the Netherlands who works with self-organisation of social systems. There are six key note lectures and fourteen sessions, representing not only the overall theme of the meeting, but the diversity of interests within ISAE as a whole. In addition, there are five workshops, posters on display for the whole conference and the opportunity to join technical tours to historical sites and applied animal research facilities in the neighbourhood of Uppsala or Stockholm.

The people who have helped organise this conference are acknowledged in these proceedings, but we would also like to acknowledge the many others who have contributed - you know who you are. On behalf of the local organising committee, we would like to welcome all participants and wish you an enjoyable and intellectually stimulating 44th Congress of the ISAE.

Lena Lidfors, Harry Blokhuis and Linda Keeling
Department of Animal Environment and Health

Acknowledgements

Main organisers

Lena Lidfors
Maria Andersson

Scientific committee

Lena Lidfors
Bo Algers
Arnd Bassler
Harry Blokhuis
Kristina Dahlborn
Jan Hultgren
Per Jensen

Linda Keeling
Per Peetz Nielsen
Nadine Reefmann
Margareta Rundgren
Elin Spangenberg
Anna Wallenbeck
Jenny Yngvesson

Organising committee

Maria Andersson
Helene Axelsson
Emma Brunberg
Madeleine Hjälm
Birgitta Larsson
Jenny Loberg
Anna Lundberg
Jenny Mattsson

Yezica Norling
Therese Rehn
Birgitte Seehuus
Malin Skog
Elin Spangenberg
Eva-Lena Svensson
Elin Weber

Sponsor group

Lena Lidfors
Christer Bergsten
Stefan Gunnarsson

Astrid Lovén-Persson
Margareta Stigson

Conference secretariat

Karin Hornay and Åsa Landqvist at Academic Conferences

Referees

Bo Algers
Arnd Bassler
Christer Bergsten
Harry Blokhuis
Alain Boissy
Xavier Boivin
Bjarne Braastad
Oliver Burman
Andy Butterworth
Elisabetta Canali
Michael Cockram
Rick D'Eath
Kristina Dahlborn
Anne Marie De Passillé
Guiseppe De Rosa
Cathy Dwyer
Sandra Edwards
Björn Forkman
Stefan Gunnarsson
Kirsten Hagen
Anders Herlin
Jan Hultgren
Margit Bak Jensen
Linda Keeling
Jörgen Kjäer
Ute Knierim
Jan Ladewig
Jan Langbein
Lena Lidfors
Jenny Loberg
Xavier Manteca

David Marlin
Lindsay Matthews
Marie-Christine Meunier Salaun
Virginie Michel
Michela Minero
Randi Moe
Lene Munksgaard
Ruth Newberry
Per Peetz Nielsen
Hans Oester
Nadine Reefmann
Bas Rodenburg
Margareta Rundgren
Jeffrey Rushen
Eva Sandberg
Heike Schulze Westerath
Elin Spangenberg
Marek Spinka
Hans Spoolder
Berry Spruijt
Franck Tuyttens
Anna Valros
Kees van Reenen
Antonio Velarde
Herman Vermeer
Eberhard von Borell
Susanne Waiblinger
Anna Wallenbeck
Beat Wechsler
Christoph Winckler
Ewa Wredle

Sponsors

www.astrazeneca.com

www.uppsala.se

LANTBRUKARNAS
RIKSFÖRBUND

www.lrf.se

www.kraiburg-agri.eu

VECTRONIC
Aerospace

www.vectronic-aerospace.com

www.ksla.se

General information

Conference venue

Uppsala Concert and Congress Hall is located in the city centre, next to the Central station. The address is Vaksala Torg 1, entrance from Vaksala Torg and from Storgatan (No. 8 on the map).

Official language

Official language of the meeting is English.

Registration and information desk (Ground floor)

Opening hours:

Tuesday August 3:	13:00-21:00
Wednesday August 4:	08:00-21:00
Thursday August 5:	08:00-12:30
Friday August 6:	08:00-17:00
Saturday August 7:	08:30-17:30

Phone: +46 (0) 730 23 84 32 (available only during secretariat opening hours)
E-mail isae.2010@slu.se

Name badges

Your name badge is your admission to the venues, scientific sessions, poster sessions and to the lunch and coffee breaks. It should be worn at all times at the conference venue and at social events.

Poster and exhibition area

The poster and exhibition area will be on the 3rd and 6th floor at Uppsala Concert and Congress Hall. We encourage you to visit the exhibition.

Internet access

Wireless Internet access is available.
Username: ukk, password: ukk

Receipt of payment and certificate of attendance

If you need a receipt of payment or a certificate of attendance, please ask for it at the information desk.

Coffee breaks

Coffee and refreshments will be served on the 6th floor.

Lunches

Lunch will be served at the ground floor.

Welcome reception, August 3rd 19:00-21:00

The welcome reception will take place at the ground floor. A buffet and drinks will be served.

Excursions, August 5th, 12:45
(Please note that pre-registration is required).

All excursion buses will depart in front of Uppsala Concert and Congress Hall, Storgatan at 12:45. Lunch is included.

Banquet at Uppsala Castle, August 6th, 19.00
(Please note that pre-registration is required).

The conference banquet will be held at Uppsala Castle, which dates back to the 16th century and has been the site of numerous historical and sometimes bloody events. The conference banquet will be a three-course gourmet dinner in the magnificent Hall of State, where Queen Christina of Sweden abdicated her throne about 350 years ago, in June of 1654 (No. 9 on the map).

Farewell party Saturday, August 7th, 19.00
(Please note that pre-registration is required).

Crayfish party at Norrlands Student Club (No. 10 on the map).

Banking service, currency

Swedish Krona (SEK) is the official currency in Sweden. An exchange office is available next to the tourist office (Forex). There are plenty of cash dispensers in Uppsala. Major international credit cards are accepted in most hotels, shops and restaurants.

Shopping in Uppsala

Most stores in Uppsala are open 10:00-19:00 on weekdays and 10:00-17:00 on Saturdays. Some stores are open on Sundays as well. Grocery stores usually have longer opening hours.

Transport from and to Stockholm Arlanda International Airport

Taxi:
You can pre book a taxi at Taxi Kurir +46 (0) 18 123 456, www.taxikurir.se or at Uppsala Taxi +46 (0) 18 100 000, www.uppsalataxi.se. The price to get to Stockholm Arlanda International Airport is about SEK 415 (45 Euro). All taxis accept credit cards.

Bus:
Bus 801 runs between Uppsala Central Station and Arlanda twice/hour from about 4 am until midnight. The journey takes about 45 minutes and costs 100 SEK (11 Euro). You buy your ticket from the driver.

Train:
Trains leave Uppsala Central Station for Arlanda Airport 1-3 times/hour from 5:00 until 23:00. The journey takes 15-20 minutes and costs 95-140 SEK (10-14 Euro) if purchased in advance at Uppsala Central Station. If you need help to get to another airport, please ask for information at the information desk.

Emergency calls

You should call 112 if anything happens which means that an ambulance, the police or the fire brigade need to be called out. 112 is a special emergency number you can call wherever you may find yourself, from a fixed or a mobile telephone.

Local conference secretariat

Academic Conferences
P.O. Box 7059,
750 07 UPPSALA, Sweden
www.akademikonferens.uu.se
tel: +46 (0) 18 67 10 03
fax: +46 (0) 18 67 35 30

Map of central Uppsala

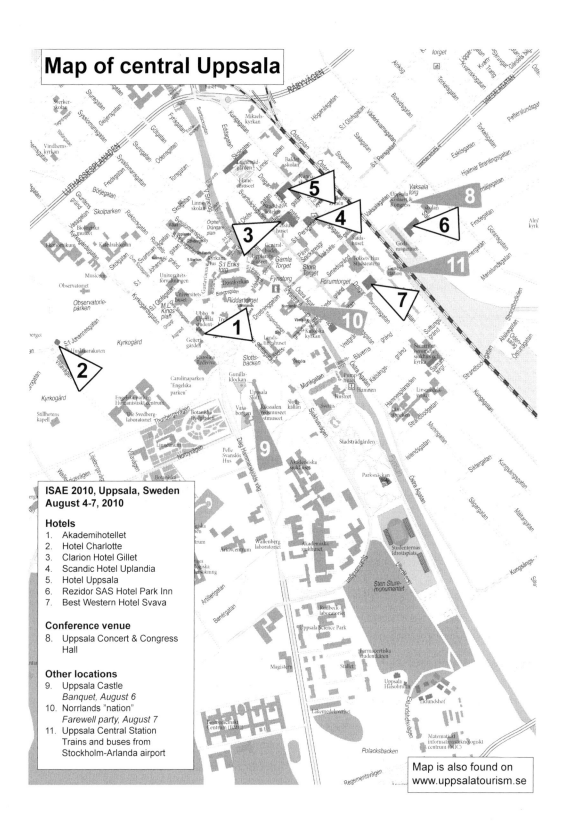

ISAE 2010, Uppsala, Sweden
August 4-7, 2010

Hotels
1. Akademihotellet
2. Hotel Charlotte
3. Clarion Hotel Gillet
4. Scandic Hotel Uplandia
5. Hotel Uppsala
6. Rezidor SAS Hotel Park Inn
7. Best Western Hotel Svava

Conference venue
8. Uppsala Concert & Congress Hall

Other locations
9. Uppsala Castle
   Banquet, August 6
10. Norrlands "nation"
   Farewell party, August 7
11. Uppsala Central Station
   Trains and buses from
   Stockholm-Arlanda airport

Map is also found on
www.uppsalatourism.se

Programme at a glance

Tuesday, August 3		
13:00 - 21:00	Registration (ground floor), posters (6th floor) & exhibitions (3rd floor)	
19:00 - 21:00	Welcome reception (ground floor)	

Wednesday, August 4	Main hall 6th floor	Hall B 3rd floor
08:00 - 09:00	Registration (ground floor)	
08:30 - 09:00	Opening ceremony	
09:00 - 10:00	Wood-Gush Memorial Lecture	
10:00 - 10:30	Coffee	
10:30 - 11:10	Key note lecture	
11:15 - 14:30	**Session 1**: Behavioural and physiological coping mechanisms in groups	**Session 2:** Cognition, learning and motivation
12:30 - 13:30	Lunch	
14:30 - 15:00	Coffee	
15:00 - 17:30	Workshops	
19:00 - 21:00	Posters, wine and cheese	

Thursday, August 5	Main hall 6th floor	Hall B 3rd floor
08:30 - 09:10	Key note lecture	
09:15 - 12:30	**Session 3:** Domestication and group behaviour	**Session 4:** Pain and sickness behaviour
10:15 - 11:00	Coffee & posters	
12:45 - 22:00	Excursions with lunch box in the bus	

Friday, August 6	Main hall 6th floor	Hall B 3rd floor
08:30 - 09:10	Key note lecture	
09:15 - 12:00	**Session 5:** Early development and maternal behaviour	**Session 6:** Human-animal interaction and welfare
10:00 - 11:00	Coffee, posters & exhibitions	
12:00 - 13:00	Lunch	
13:00 - 13:40	Key note lecture	
13:45 - 16:30	**Session 7:** Cognition and emotions in social contexts	**Session 8:** Housing and outdoor access
14:30 - 15:00	Coffee & posters	
16:30 - 18:00		ISAE members annual meeting
19:00 - 01:00	Banquet at Uppsala Castle	

Saturday, August 7	Main hall 6th floor	Hall B 3rd floor
09:00 - 09:40	Key note lecture	
09:45 - 12:00	**Session 9:** Social adaptations to large groups	**Session 10:** Behaviour and genetics
10:30 - 11:00	Coffee & posters	
12:00 - 13:00	Lunch	
13:00 - 13:40	Key note lecture	
13:45 - 14:45	**Session 11:** Abnormal behaviour	**Session 12:** Feeding behaviour
14:45 - 15:45	Coffee & posters	
15:45 - 16:30	**Session 13:** Male reproduction	**Session 14:** Free papers
16:30 - 17:00	Closing of congress	
19:00	Farewell Party at Norrlands Nation	

Programme for ISAE Congress

All activities will occur in Uppsala Concert & Congress if not stated otherwise.
Only the first author of the abstract is named in the programme, but in some cases the paper will be presented by a co-author.

Tuesday, August 3

09:00 - 17:00	Council meeting (Room K2)
10:00 - 17:00	EFBA-meeting (Room K7)
10:00 - 15:00	Preparation for posters, exhibitions and registrations
12:00 - 13:00	**Lunch (ground floor)**
13:00 - 21:00	Registration (ground floor), posters (6th floor) & exhibitions (3rd floor)
19:00 - 21:00	Welcome reception (ground floor)
	Dean of Faculty welcome (Kerstin Svennersten-Sjaunja)
	Organisers welcome (Lena Lidfors, Maria Andersson)

Wednesday, August 4

08:00-09:00	Registration (ground floor)
08:30-09:00	Opening ceremony
	Organising committee - Lena Lidfors & Maria Andersson
	Deputy Vice-Chancellor - Lena Andersson-Eklund
	(Main hall 6th floor)
09:00-10:00	Wood-Gush Memorial Lecture - Charlotte Hemelrijk
	'Self-organisation of social systems of animals'
	(Main hall 6th floor)
10:00-10:30	**Coffee**
10:30-11:10	Key note lecture - Jaap Koolhaas
	'The adaptive and maladaptive nature of the social stress response'
	(Main hall 6th floor)

11:15-12:30	Parallel sessions	
	Main hall 6th floor	**Hall B 3rd floor**
	Session 1: Behavioural and physiological coping mechanisms in groups	*Session 2:* Cognition, learning and motivation
11:15	Bas Rodenburg 'Maternal care reduces cannibalistic toe pecking in domestic laying hens'	Elise Gieling 'Performance of low and normal birth weight piglets in a cognitive holeboard task'
11:30	Simone Helmreich 'Influence of night-time milking and lying on nocturnal salivary melatonin concentration of dairy cows on automatic milking farms'	Randi Moe 'Dopamine regulation of reward-related behaviours in laying hens'
11:45	Christine Nicol 'Relationships between multiple behavioural and physiological indicators taken from group and individually-housed chickens'	Janne Winther Christensen 'Object habituation in horses'
12:00	Anke Gutmann 'Heart rate variability during allogrooming in dairy cows: a pilot study'	Sabrina Mueller 'How do pigs react to situations of different emotional valence?'
12:15	Elizabeth Bolhuis 'Effects of environmental enrichment and loose housing of sows on behaviour and growth of weanling piglets'	Stephanie Buijs 'Will work for space: broiler chickens are motivated to move to lower stocking densities'
12:30-13:30	**Lunch (ground floor)**	
13:30-14:30	Continued parallel sessions	
	Main hall 6th floor	**Hall B 3rd floor**
13:30	Emily O'connor 'The effect of environmental stressors on the social behaviour and stress physiology of group-housed growing pigs'	Anette Wichman 'A happy hen? Comparing two different measures of emotional state'
13:45	Jitka Šilerová 'Sow nursing behaviour in isolation and in acoustic and visual contact'	Oliver Burman 'Comparing the influence of pre and post-consumption phases of a rewarding event on cognitive bias in dogs'
14:00	Eva Nordmann 'Feed barrier design affects behaviour and adrenocortical activity of goats'	Lisa Tomkins 'A measure of sensory (visual) lateralization in dogs (Canis familiaris): the Sensory Jump Test'
14:15	Laura Webb 'Comparing the effects of different roughage diets on grooming behaviour and hairball prevalence in veal calves'	Jasmin Kirchner 'Calling individual sows to feed: a possibility to reduce agonistic interactions in electronic sow feeding systems?'
14:30-15:00	**Coffee**	
15:00-17:30	Workshops	
19:00-21:00	**Posters, wine and cheese**	

Thursday, August 5

08:30-09:10	Key note lecture- Marek Spinka 'Emotional signalization, domestication and animal welfare' (Main hall 6th floor)	
09:15-10:15	**Parallel sessions** **Main hall 6th floor**	**Hall B 3rd floor**
	Session 3: Domestication and group behaviour	*Session 4:* Pain and sickness behaviour
09:15	Somparn Pipat 'Provision of designated defecation areas to reduce the spatial distribution of faeces by grazing swamp buffalo heifers'	Ann-Helena Hokkanen 'Five-day headache after disbudding?'
09:30	Peter Krawczel 'Short-term overcrowding affects the behavior of lactating Holstein dairy cows negatively'	Gina Caplen 'Altered thermal but not mechanical nociceptive threshold in moderately lame broiler chickens'
09:45	Beatrix Eklund 'Domestication effects on behavioural synchronization in chickens'	Margit Bak Jensen 'Behavioural responses to disease in dairy calves'
10:00	Lucy Asher 'Flocking for food or flockmates? A model of chicken social behaviour'	Lene Mølgaard 'Effects of sampling liver biopsies on dairy cow behaviour'
10:15-11:00	**Coffee & posters**	
11:00-12:30	**Continued parallel sessions** **Main hall 6th floor**	**Hall B 3rd floor**
11:00	Markus Jöngren 'Selection for reduced fear of humans in red junglefowl: correlated effects on behaviour and production traits'	Katrine Fogsgaard Jensen 'Sickness behaviour in dairy cows challenged with *E. coli* induced mastitis'
11:15	Siobhan Abeyesinghe 'Do hens have friends?'	Jutta Berk 'Relationship between forced physical activity, leg health and performance in heavy male turkeys'
11:30	Thorsten Pickel 'Perch material and diameter affects perching behaviour in laying hens'	Kristen Walker 'Behavioural responses of juvenile Steller sea lions (Eumetopias jubatus) to hot-iron branding'
11:45	Candace Croney 'Effects of early rearing environment on learning ability and behavior of laying hens'	Maria Ylä-Ajos 'Use of a runway test in demonstrating turkey muscle pain'
12:00	Kristi Bovey 'The effect of birth weight and age on the behavioural responses of piglets to tail docking and ear notching'	Sandra Düpjan 'Domestic pigs' responses to conspecific stress-induced vocalisation'
12:15	Maja Makagon 'Nest box sharing by Pekin ducks: a social network analysis'	
12:45-22:00	**Excursions with lunch box in the bus**	

Friday, August 6

08:30-09:10	Key note lecture- Hugo Lagercrantz 'The emergence of human consciousness: from fetal to neonatal life' (Main hall 6th floor)	
09:15-10:00	**Parallel sessions**	
	Main hall 6th floor	**Hall B 3rd floor**
	Session 5: Early development and maternal behaviour	*Session 6:* Human-animal interaction and welfare
09:15	Cathy Dwyer 'Undernutrition in early to mid pregnancy causes deficits in the expression of maternal behaviour in sheep that may affect lamb survival'	Ines Windschnurer 'Modifying attitudes and behaviour towards dairy cattle by multimedia-based cognitive-behavioural intervention'
09:30	Eva Mainau 'Effect of time and parity on the behaviour of dairy cows during the puerperal period'	Xavier Boivin 'Environmental factors influencing responses of calves to human contact'
09:45	Alice Barrier 'Calving difficulty in dairy cows results in less vigorous calves but does not affect the onset of maternal behaviour'	Therese Rehn 'Investigating greeting behaviour in dogs reunited with a familiar person'
10:00-11:00	**Coffee, posters & exhibitions**	
11:00-12:00	**Continued parallel sessions**	
	Main hall 6th floor	**Hall B 3rd floor**
11:00	Gudrun Illmann 'Non-nutritive nursings: one mechanism to control milk transfer in domestic pigs'	Rebecca Meagher 'When inactivity predicts poor reproductive success in farmed mink, is poor welfare the link?'
11:15	Anna Gudrun Thorhallsdottir 'Comparison of growth and performance of dairy calves suckling mother, suckling foster mother, fed ad lib milk replacer and fed conventionally'	Nancy Clarke 'Belief in animal sentience during veterinary education'
11:30	Marije Oostindjer 'Effects of maternal presence and environmental enrichment on food neophobia of piglets'	Stephanie Sinclair 'A comparison of the effects of three dehorning methods on the welfare of Bos indicus calves'
11:45	Sophie Hild 'Effects of positive versus negative handling during pregnancy on maternal behaviour in sheep'	Mariëlle R.N. Bruijnis 'Welfare impact of foot disorders in dairy cattle'
12:00-13:00	**Lunch (ground floor)**	

13:00-13:40	Key note lecture- Mauricio Papini	
	'Social consequences of surprising incentive devaluations'	
	(Main hall 6th floor)	

13:45-14:30	**Parallel sessions**	
	Main hall 6th floor	**Hall B 3rd floor**
	Session 7: Cognition and emotions in social contexts	*Session 8:* Housing and outdoor access
13:45	Ragen T.S. Mcgowan	Cassandra Tucker
	'Positive affect and learning: exploring the 'Eureka Effect' in dogs'	'Muddy conditions reduce lying time in dairy cattle'
14:00	Manuela Zebunke	Gemma Charlton
	'Coping with cognitive challenge is emotionally positively evaluated by pig'	'The effect of TMR on dairy cow preference to be indoors or at pasture'
14:15	Samantha Jones	Erin Mintline
	'Cage-mate/non-cage-mate categorisation of social odours in rats'	'Both pen size and shape affect locomotor play in dairy calves'
14:30-15:00	**Coffee & posters**	
15:00-16:15	**Continued parallel sessions**	
	Main hall 6th floor	**Hall B 3rd floor**
15:00	Nadine Reefmann	Gudrun Plesch
	'Does human handling affect cognitive bias in dogs?'	'Lying down duration in German Holstein dairy cows housed in cubicles in relation to body size and cubicle characteristics'
15:15	Rebecca Doyle	Karin Keckeis
	'Administration of serotonin inhibitor p-Chlorophenylalanine induces pessimistic-like judgement bias in sheep'	'Behaviour of different dairy cattle genotypes in pasture-based production systems'
15:30	Birgitte Seehuus	Marianna Norring
	'Investigating the reward cycle in chickens'	'Milk yield affects time budget in dairy cows'
15:45	Corinna Clark	Hrefna Sigurjónsdóttir
	'Early-life-experiences alter attention to potential threats in sheep'	'Interactions between herd keeping stallions in a big enclosure'
16:00	Jan Langbein	
	'Cognitive abilities of farm animals: 'Stupid goats' (Capra hircus) are able to form open-end categories'	
16:30-18:00		ISAE members annual meeting
19:00-01:00	**Banquet at Uppsala Castle**	

Saturday, August 7

09:00-09:40	Key note lecture- Maja Makagong 'How can social network analysis contribute to social behaviour research in applied ethology?' (Main hall 6th floor)	
09:45-10:30	**Parallel sessions** **Main hall 6th floor**	**Hall B 3rd floor**
	Session 9: Social adaptations to large groups	*Session 10:* Behaviour and genetics
09:45	Radka Sarova 'Graded leadership of dominant beef cows: stronger in space than in time?'	Emma Brunberg 'Differences in hypothalamic gene expression in feather pecking chickens'
10:00	Jonathan Cooper 'Effect of tree cover on ranging behaviour of free range hens'	Anna-Carin Karlsson 'The pigmentation regulating PMEL17 gene has pleiotropic effects on behavior in chickens'
10:15	Anna Wallenbeck 'Predicting tail biting outbreaks among growing-finishing pigs under commercial conditions'	Björn Forkman 'Cuteness and playfulness in dogs, are they related?'
10:30-11:00	**Coffee & posters**	
11:00-12:00	**Continued parallel sessions** **Main hall 6th floor**	**Hall B 3rd floor**
11:00	Taylor Donnalee 'The effect of flock size and paddock complexity on following behaviour in Merino sheep'	Stephanie Matheson 'Heritabilities of birth assistance and neonatal lamb traits in Suffolk sheep'
11:15	Isabelle Castro 'Social behaviour during winter outdoor exercise of cows of the Hérens breed kept in tie-stalls'	Daniel Nätt 'Transgenerational epigenetic inheritance of behavioural traits in the chicken (Gallus gallus)'
11:30	Elke Hartmann 'Effects of three methods for mixing unfamiliar horses on their potential to minimise aggressive interactions'	Linda Rosager Duve 'Social bonding and sociability are affected by the level of social contact in dairy calves'
11:45	Sabine Gebhardt-Henrich 'Use of outdoor range in large groups of laying hens'	Hans Spoolder 'Space requirements for group housed pigs at high ambient temperatures'
12:00-13:00	**Lunch (ground floor)**	

13:00-13:40	Key note lecture- Georgia Mason	
	'Why do captive animals perform repetitive abnormal behaviour?'	
	(Main hall 6th floor)	

13:45-14:45	**Parallel sessions**	
	Main hall 6th floor	**Hall B 3rd floor**
	Session 11: Abnormal behaviour	*Session 12:* Feeding behaviour
13:45	Megan Anne Jones	Anne Marie De Passillé
	'Why wild-caught animals seldom stereotype'	'Which behaviour of dairy calves best indicates their level of hunger during weaning?'
14:00	Jamie Dallaire	Kristina Dahlborn
	'Recurrent perseveration correlates with cage stereotypic behaviour and is insensitive to environmental enrichment in American mink, Mustela vison'	'Shade seeking, feeding behaviour and social interactions in lactating camels (Camelus Dromedarius) on four different watering regimes'
14:15	Camilla Munsterhjelm	Carol Souza Da Silva
	'Health in tail biters and bitten pigs in a matched case-control study'	'Screening the satiating properties of dietary fibre sources in adult pigs'
14:30	Karen Thodberg	Sunil Kumar Pal
	'The risk of tail-biting in relation to level of tail-docking'	'Inter-group aggressive behaviour of free-ranging dogs (Canis familiaris)'

14:45-15:45	**Coffee & posters**	

15:45-16:30	**Continued parallel sessions**	
	Main hall 6th floor	**Hall B 3rd floor**
	Session 13: Male reproduction	*Session 14*: Free paper
	(Main hall 6th floor)	(Hall B 3rd floor)
	Maria Díez-León	Hannah Salvin
15:45	'Male mink (Mustela vison) from enriched cages are more successful as mates'	'Important considerations for the use of older adult dogs in ethological research'
16:00	Herman Vermeer	Nadine Ringgenberg
	'Management measures to reduce sexual behaviour in male pigs'	'Validation of accelerometers to automatically record sow postures and stepping behaviour'
16:15		J. Mas Muñoz
		'Variation in growth, feed intake and behaviour of sole'

16:30-17:00	**Closing of congress**	
19:00	Farewell Party at Norrlands Nation	

Workshops
To be held in rooms K1, K2, K3, K4 and Sal C.

1. Animal welfare risk assessment

Organisers: Jan Hultgren and Bo Algers
Animal welfare risks need to be assessed at national and global levels in response to consumer concern and legislation. Guidelines on animal welfare RA are being developed by the European Food Safety Authority. So far, a number of limitations and problems have been identified, relating to the specific nature of animal welfare. The workshop will address different aspects of the applicability of RA methods to animal welfare.

Programme

Moderator: Jörg Hartung
15:00 - 15:15 Welcome, presentation of contributors, practicalities. Jan Hultgren
15:15 - 15:30 **Animal Welfare Risk Assessment – stating the problem**. Riitta Maijala
15:30 - 15:45 **Expert elicitation in animal welfare risk assessment – methods and limitations**. Marc Bracke
15:45 - 16:00 **Animal Welfare Risk Assessment in national official control – a practical approach**. Jan Hultgren and Bo Algers
16:00 - 16:45 Exercise
16:45 - 17:15 General discussion
17:15 - 17:30 Summary and conclusions. Oriol Ribó Arboledas

2. Welfare monitoring using automatic measurement of behaviour

Organiser: Margit Bak Jensen
Sensor technology offers new possibilities for monitoring behaviour in relation to welfare, e.g. identification of diseased individuals in large groups and surveillance around parturition. The workshop will give a short overview of on-line measures of behaviour in cattle and swine, including the applied technology and modelling. The workshop will discuss perspectives of ongoing research and future possibilities for on-line behaviour surveillance to improve welfare in modern livestock farming.

Programme

15:00 - 17.30 **Technical prospects for automated behavioural monitoring.** Matti Pastell
Automated farrowing prediction and birth surveillance in farm animals. Lene Juul Pedersen
Automated detection of lameness in cows. Jeff Rushen
Using feeding behaviour to assess feed bunk access, competition and illness in dairy cattle. Trevor DeVries

3. Perinatal experiences and animal welfare

Organiser: Per Jensen
This workshop brings together research from different species, and deals with effects on behaviour and welfare of experiences encountered from the point of conception until shortly after birth. This is an immensely growing area of science, and integrates methods and concepts from ethology, stress physiology, and genomics.

Programme

15:00 - 15:15 Introduction to the workshop. Per Jensen
15:15 - 15:30 **Maternal effects on the welfare of chickens and fish.** Andrew Janczak
15:30 - 15:45 **How perinatal and early-life experiences affect behavioural development in domestic chicks.** Bas Rodenburg
15:45 - 16:00 **Prenatal life and animal welfare in the domestic pig.** Kenny Rutherford
16:00 - 16:15 Content related questions and discussion
16:15 - 17:15 General discussion. Chaired by Per Jensen and Liesbeth Bolhuis
17:15 - 17:30 Summary and conclusions. Liesbeth Bolhuis

4. Are perches responsible for keel bone deformities?

Organiser: Sabine G. Gebhardt-Henrich
What role is behaviour (crashes and sitting on perches) likely to play in the occurrence of keel bone fractures and deformities and what kind of behavior (crashes, egg-laying) is mainly responsible?
How could behavior be modified to decrease the frequency of fractures and deformities, by aviary and cage design and/or management?
Do (recent, unhealed) fractures and deformations affect the behavior of laying hens, i.e. how is the welfare of the affected birds?

Programme

15:00 - 15:05 Introduction. S.G. Gebhardt-Henrich
15:05 - 15:20 **Prevalence of keel bone deformities with an emphasis of Swiss laying hens** S. Käppeli, S.G. Gebhardt-Henrich, A. Pfulg, E. Fröhlich, M.H. Stoffel
15:20 - 15:35 **Perching Behaviour and Pressure Load on Keel Bone and Foot Pads in Laying Hens** B. Scholz, T. Pickel, L. Schrader
15:35 - 15:50 **Factors affecting the movement between perches in laying hens** M.J. Haskell, C. Moinard, V. Sandilands, K.M.D. Rutherford, P. Statham, S. Wilson, P. R. Green
15:50 - 16:05 **Are perches responsible for keel bone deformities in laying hens?** V. Sandilands, L. Baker, S. Brocklehurst, L. Toma, C. Moinard
16:05 - 16:20 **Influence of access to aerial perches on keel bone injuries in laying hens on commercial free range farms** C.J. Nicholson, N.E. O'Connell
16:20 - 16:50 Work in groups, cards on pin board everybody
16:50 - 17:20 Discussion everybody
17:20 - 17:30 Summary. S.G. Gebhardt-Henrich

5. The animal welfare conservation interface: applied ethology and human interventions in wild animal populations

Organisers: Chris Draper and Pete Goddard
Species conservation, from research in the field to wildlife management actions, may impact on the welfare of individual animals. Conservation practitioners and animal welfare scientists may differ in their perceptions of the importance of animal welfare versus species conservation, and in their understanding of the science of these two increasingly emergent fields. This workshop will discuss the broad areas of conflict or benefit between animal welfare and conservation, with a view to identifying underlying principles for further attention, while assessing the perceptions of ISAE conference participants and members.

Programme

Moderator: Chris Draper, Born Free Foundation, University of Bristol, UK
The workshop will take the form of five 10-minute presentations, with introductory overview and summing up, followed by a group discussion and moderated Q&A session.

15:00 - 17.30 Introduction and overview. Chris Draper, Born Free Foundation / University of Bristol, UK

Conservation in cages: captive animal welfare and *ex situ* conservation. Chris Draper

Reintroduction success: should this be planned at the individual or species level? Samantha Bremner-Harrison

Wild animals as pets: animal welfare and conservation implications. Paul Koene

The welfare of extensively managed animals: an interface between captive animals and wildlife. Cathy Dwyer

A welfare construct for wild animals: can it mean anything to them? Pete Goddard

Group discussion and moderated Q&A session
Summary. Pete Goddard

Abstracts

Session 2. Cognition, learning and motivation

Session 3. Domestication and group behaviour

Session 4. Pain and sickness behaviour

Session 5. Early development and maternal behaviour

Session 6. Human animal interaction and welfare

Session 7. Cognition and emotions in social contexts

Session 8. Housing and outdoor access

Session 9. Social adaptations to large groups

Session 10. Behaviour and genetics

Session 11. Abnormal behaviour

Session 12. Feeding behaviour

Session 13. Male reproduction

Session 14. Free papers

Poster session

Workshop 1. Animal welfare risk assessment

Workshop 2. Welfare monitoring using automatic measurement of behaviour

Workshop 3. Perinatal experiences and animal welfare

Workshop 4. Are perches responsible for keel bone deformities?

Workshop 5. The animal welfare and conservation interface: applied ethology and human interventions in wild animal populations

David Wood-Gush Memorial Lecture

Self-organisation of social systems of animals

Hemelrijk, Charlotte, RUG, Kerklaan 30, 9751NN Haren, Netherlands; c.k.hemelrijk@rug.nl

Individual based models with a high potential for self-organisation have shown that cognitively simple rules in individuals may lead to complex collective patterns. I will illustrate this with examples of moving groups of animals, i.e. fish, birds and primates. As regards schools of fish, we have shown that, unexpectedly, certain traits emerge, that are considered to be protective against predation: schools are oblong with the highest density at the front. These traits arise as a side-effect of slowing down to avoid collision among individuals that coordinate with each other. We find support for similar processes in schools of mullets. As regards starlings, empirical data show that their flock shape is highly variable. We get similar flocking patterns in a model by extending that of coordinating individuals with flying behaviour and attraction to a sleeping site. The banking by individuals while turning appears essential. Primate social organisation is complex. However, we show that a model (called DomWorld) and its extension (GroofiWorld) deliver patterns of aggression and affiliation (mainly by grooming) that resemble egalitarian and despotic species of macaques in many aspects despite the low level of cognition in the model: In DomWorld, individuals merely group and compete and in GroofiWorld, they also groom others if they think that they will loose a fight. For instance, in the absence of a motivation to reconcile, patterns of reconciliation emerge similar to those of primates. Further, the prediction by DomWorld that by increasing intensity of aggression (from 'threats' to 'bites') female dominance over males increases, is confirmed in empirical data of primates. These kinds of models help us to develop new hypotheses about the integration of different traits. They lead to a different view about the cognition involved in, and the evolution of, these types of societies.

The adaptive and maladaptive nature of the social stress response

Koolhaas, Jaap, University Groningen, Dept. Of Behavioral Physiology, P.O. Box 14, 9750 AA Haren, Netherlands; j.m.koolhaas@rug.nl

With the steadily increasing number of publications in social stress research it has become evident that the conventional usage of the stress concept bears considerable problems. The inflated use of the term 'stress' to conditions ranging from even the mildest challenging stimulation to severely aversive conditions is inappropriate. The social environment is not only an important source of social stress requiring strong adaptations it also buffers against the adverse effects of social stress. This dual nature asks for a clear delineation of the adaptive role of the stress response from its maladaptive actions. Review of the literature reveals that the physiological 'stress' response to positive, rewarding social stimuli that are often not considered as stressors can be as large as the response to negative and aversive social stimuli. Based on a more detailed analysis of the neuroendocrine responses to social stimuli, we conclude that it is not the magnitude of the HPA axis or sympathetic response that dissociates a stimulus from a stressor but rather the downward slope of these responses (controllability) and the anticipatory response (predictability). Analysis of the physiological response during exercise supports the view that the magnitude of the neuroendocrine response reflects the metabolic and physiological demands required for behavioural activity. This more refined view has consequences for the interpretation of results in terms of the adaptive and/or maladaptive nature of the response and hence for the interpretation in terms of animal welfare. Indeed, studies aimed at the individual variation in the consequences of social subordination show considerable changes in behavior and in several parameters of the HPA axis in reactive coping males in particular. These changes are generally interpreted as the first signs of stress pathology and reduced welfare. However, these changes may just as well reflect the process of adaptation. These two alternative views will be discussed on the basis of evidence that coping styles differentially rely on predictability and controllability.

Emotional signalization, domestication and animal welfare

Špinka, Marek, Institute of Animal Science, Department of Ethology, Pratelstvi 815, 104 00 Prague - Uhrineves, Czech Republic; spinka@vuzv.cz

Animals may transfer emotions among themselves, with consequences for their welfare. The process called emotional contagion causes animals to shift, upon perceiving animals in an emotional state, their own affective state in the same direction. Emotional contagion may work through automatic mimicking the perceived expressions which in turn affects the emotional state of the mimicking animal. This process can have an avalanche character in animal groups and has been observed, for instance, in play and fear in pigs. Surprisingly, very little quantitative data exist about this important phenomenon. For humans, emotionally loaded animal signals can be windows to assess and tools to influence emotional states of animals. Humans also perceive emotional signals from animals. Thus, beside material goods and services, humans have acquired emotional benefits from domestic animals such as perceiving that the animals are content and happy in their care. Therefore humans may have been, during domestication, promoting positive emotional (PE) signals from the animals. This can be achieved in four ways that will be exemplified on dogs. (1) We can improve captive environments so that animals more often experience PE states; through selecting animals that emit most PE signals, we can unintentionally change the genetic and epigenetic makeup of animals so that they (2) experience PE states even in not-so-good environments and/or (3) emit PE signals even when their emotional state is not positive. Finally, we can (4) auto-adjust ourselves to overestimate the positive signals received from domesticated animals. Mirror wise, with negative emotions (NE), analogous processes may lead us to removing challenges from the environment that cause NE states, suppressing the animals' capacity to feel NE states, decreasing their inclination to express NE states through NE signals, and compromising our ability to perceive the NE states emitted by the animals. This possibility will be illustrated on dairy cows. Emotions include an important social dimension. Paying attention the social dimension of emotions, both within species of animals whose life we manage, and between animals and humans, will promote our understanding of animal welfare and may open new ways to affect it positively.

The emergence of human consciousness: from fetal to neonatal life
Lagercrantz, Hugo, Karolinska Institute, Astrid Lindgren Children´s Hospital, Stockholm, Sweden; Hugo.Lagercrantz@ki.se

A simple definition of consciousness is sensory awareness of the body, the self, and the world. The foetus may be aware of the body, for example by perceiving pain. It reacts to touch, smell, and sound, and shows facial expressions responding to external stimuli. However, these reactions are probably pre-programmed and have a subcortical nonconscious origin. Furthermore, the foetus is almost continuously asleep and unconscious partially due to endogenous sedation. Conversely, the newborn infant can be awake, exhibit sensory awareness, and process memorized mental representations. It is also able to differentiate between self and non-self touch, express emotions, and show signs of shared feelings. Yet, it is unreflective, present oriented, and makes little reference to concept of him/herself. Newborn infants display features characteristic of what may be referred to as basic consciousness and they still have to undergo considerable maturation to reach the level of adult consciousness. The preterm infant, *ex utero*, may open its eyes and establish minimal eye contact with its mother. It also shows avoidance reactions to harmful stimuli. However, the thalamocortical connections are not yet fully established, which is why it can only reach a minimal level of consciousness.

Social consequences of surprising incentive devaluations

Papini, Mauricio R., Texas Christian University, Department of Psychology, Fort Worth, TX 76129, USA; m.papini@tcu.edu

Animals exposed to surprising incentive devaluations exhibit a variety of behavioral and physiological reactions that can be categorized as frustrating. Frustration is defined as the internal emotional state induced by surprising incentive devaluations or omissions. For example, consummatory behavior is suppressed, instrumental behavior is redirected, dominant responses are invigorated, and pain sensitivity is reduced. Some of these effects depend on neurochemical systems known to be involved in anxiety and reward processes (e.g., GABA and opioid receptors), induce the release of stress hormones, and are modulated by activity in limbic areas of the brain (e.g., amygdala and anterior cingulate cortex). Incentive devaluations also affect social behaviors. Three areas of research will be reviewed in this presentation. (a) Agonistic behavior. Classic studies with pigeons, rats, and monkeys have shown that surprising food omissions trigger aggressive behavior toward live or stuffed conspecifics. Recent research shows that such omissions can also inhibit aggressive behavior in dominant male rats, thus leading to a behavioral effect analogous to learned helplessness. (b) Sexual behavior. Surprising incentive omissions also reduce sexual behavior in male rats. Interestingly, successful copulatory behavior has an anxiolytic-like effect in the incentive devaluation situation. (c) Infant vocalizations. Human and rat infants exposed to surprising incentive devaluations exhibit distress vocalizations typically used to induce maternal care. Recent data show that rats selected for fast re-adjustment to an episode of incentive devaluation also exhibit, as 10-day-old infants, vocalization parameters that maximize maternal care in a mother-infant separation test. The potential mediation of some of these emotional effects by differential opioid receptor efficacy will be briefly discussed. Thus, incentive devaluations both affect and are affected by social behaviors.

How can social network analysis contribute to social behavior research in applied ethology?

Makagon, Maja M. and Mench, Joy A., University of California, Davis, Animal Behavior Graduate Group and Department of Animal Science, One Shields Avenue, Davis, CA, 95616, USA; mmmakagon@ucdavis.edu

Social network (SN) metrics are increasingly used by behavioral ecologists to describe the patterns and quality of interactions among individuals. We provide an overview of this methodology, with examples illustrating how it can be used to study social behavior qualitatively in applied contexts. Like most social interaction analyses, it provides information about direct relationships (e.g. dominant-subordinate relationships). However, it also generates a more global model of social organization that determines how individual patterns of social interaction relate to individual and group characteristics. Individuals are represented as points on a graph (nodes) connected by lines (edges) which represent their interactions. Rather than concentrating on the individual nodes, SN analysis focuses on the edges and the distribution of nodes within the network. Thus, SN can not only be used to identify socially dominant individuals in the group, but also individuals who play key roles in maintaining group stability as a result of their position within the network. The SN approach also provides information about the importance of indirect connections. For example, the extent of the spread of a socially transmitted abnormal behavior like feather pecking could be predicted by ascertaining the number of group members interacting with the focal animal (i.e., the feather pecker), and the number of individuals interacting with those group members. SN analyses can also be used to quantify differences in social structure across contexts, for example by evaluating how social structures form in different environments such as housing systems that vary in terms of group size or composition. While SN metrics are most effective when evaluating interactions between individually identifiable animals that interact regularly, they can also be utilized in situations where interactions are sporadic and full group membership cannot be defined, making them useful for assessing social structures of animals housed in large groups.

Why do captive animals perform repetitive abnormal behaviour?

Mason, Georgia, University of Guelph, Department of Animal Science, Guelph, ON, Canada; gmason@uoguelph.ca

Over 85 million animals worldwide, on farms and in research labs, zoos, stables and people's homes, perform abnormal, stereotypic behaviours (SBs) such as repetitive route-tracing, or harmful plucking and biting. Ethological explanations for SBs have traditionally been in terms of frustrated motivation: normal, highly motivated behaviours are argued to occur in vacuum or redirected forms when performing them more naturalistically is impossible. In some instances, SBs are further hypothesized to help animals cope better with poor housing conditions, by providing some of the motivational feedback that normal behaviour would yield. Ethological accounts, however, are rather poor at explaining the extreme degree of repetition, or the self-harm, involved in some forms of SB. They also ignore potential explanations arising from neuroscience and psychology; fields which account for abnormal repetitive behaviours in clinical subjects (e.g. humans with autism) in terms of forebrain dysfunction, especially of neural loops running between basal ganglia and cortex. Research findings from neuroscience/ psychology on humans and manipulated 'animal models' (e.g. rodents or primates with stimulant-induced SBs), have successfully suggested useful post-mortem indices to investigate in animals that are stereotypic because held captive. Techniques derived from tests for screening humans patients also yield novel non-invasive means of assessing altered cognitive processes in live animals, and changes in how their behaviour is controlled. Studies using this approach consistently find links between captivity-induced SB and a more generalized tendency to repeat and persist with a wide range of functionless behaviours (e.g. persistent working for rewards in extinction). In many cases, such data show that impoverished captive conditions profoundly alter how the brain controls behaviour. In others, they suggest instead individual risk factors related to normal individual differences in tendencies to form routines. I will end by discussing how ethological and neuroscience-based explanations complement each other.

Maternal care reduces cannibalistic toe pecking in domestic laying hens
Rodenburg, T. Bas, Wageningen University, Animal Breeding and Genomics Centre, P.O. Box 338, 6700 AH Wageningen, Netherlands; bas.rodenburg@wur.nl

In commercial laying hen husbandry, chicks are reared in large groups without maternal care. This can result in difficulties in coping with fear and stress and increased levels of feather pecking and cannibalism. The aim of this study was to investigate the effects of brooding by a hen on behavioural development and fearfulness of the chicks and on cannibalistic pecking, when these birds are adult. Thirty-eight groups of 13 chicks each were brooded artificially and housed either with (18 groups) or without (20 groups) a foster mother (silky hen) in floor pens. At 7 wk of age, the foster mothers were removed. Two genotypes were used, but this paper will focus on the effects of brooding so means over both genotypes will be presented. Observations included ground pecking behaviour from 1-3 wk of age, open-field response at 5 wk of age and incidence of cannibalistic toe wounds at 40 wk of age. Data were analyzed using ANOVA, testing effects of mother, genotype and their interaction (with group as experimental unit). Brooded chicks showed more ground pecking (mean 133 vs. 108 bouts/observation; $F(1,37)=6.90$, $P<0.05$) than non-brooded chicks from 1 to 3 wk of age. Brooded chicks were also more active in the open-field test at 5 wk of age (63 vs. 36 steps; $F(1,37)=5.76$, $P<0.05$), indicating that they were less fearful. At 40 wk of age, the incidence of toe wounds, a major problem in this experiment, was lower in brooded birds than in non-brooded birds (15 vs. 28% birds; $F(1,37)=7.71$, $P<0.01$). These results show that maternal care favourably affects behavioural development and the ability to cope with fear and stress in chicks. It may even help to reduce cannibalism in adult laying hens. Future research should aim at translating the positive effects of maternal care to practical rearing systems.

Influence of night-time milking and lying on nocturnal salivary melatonin concentration of dairy cows on automatic milking farms
Helmreich, Simone, Gygax, Lorenz, Wechsler, Beat and Hauser, Rudolf, Centre for Proper Housing of Ruminants and Pigs, Federal Veterinary Office, Research Station Agroscope Reckenholz-Tänikon ART, 8356 Ettenhausen, Switzerland; simone.helmreich@art.admin.ch

Automatic milking enables cows to choose their individual milking times and intervals throughout 24 hours. Consequently cows also visit the milking stall at night, resulting in an interruption of their nightly lying periods. The duration of stay in an illuminated barn sector like the milking stall, may have an effect on the nocturnal melatonin concentration, which in turn may affect the quality of sleep and cow welfare. The study aimed at assessing to what extent night-time milking and lying coincide with changes in the nocturnal melatonin concentration. The experiment was conducted on eight commercial farms with automatic milking in Switzerland. 112 cows (14-16/farm) which belonged to the herd at least 5 weeks and with a low (median 6/14d) and high (median 11/14d) number of night-time milkings (22pm-5am) were selected prior to the experiment on the basis of 14-day herd management data. Individual lying times were recorded using data loggers attached to the leg of the cows that continuously logged standing or lying at intervals of 30 s. A total of ten saliva samples per cow were taken in two periods between 12am-7am on the last days of the experiment. Melatonin concentration was analysed by radioimmunoassay and the levels of nocturnal samples were averaged per cow. Average lying time per night was calculated based on data of three days without experimental intervention and the number of night-time milkings was counted during 14 days including the saliva sampling period. Data were analysed using linear mixed-effects models. Cows with a high number of night-time milkings tended to have lower nocturnal melatonin concentrations ($F_{1,101}=3.16$, p=0.08). Night-time lying duration did not covary with nocturnal melatonin concentrations ($F_{1,101}=0.72$, P=0.40). Whereas reduced melatonin concentrations indeed tended to coincide with repeated nightly exposure to light, lying time of the cows was not concurrently restricted.

Relationships between multiple behavioural and physiological indicators taken from group and individually-housed chickens

Nicol, Christine, Caplen, Gina, Edgar, Jo, Richards, Gemma and Browne, William, University of Bristol, Clinical Veterinary Science, Langford House, Langford, BS40 5DU, United Kingdom; c.j.nicol@bris.ac.uk

The assessment of coping mechanisms for group-housed animals necessarily requires the collection of multiple indicators of welfare. The aim of this study was to quantify associations between these indicators to help towards the development of more efficient welfare assessment protocols. Previous studies have examined correlations between a small number of different measures but few have (1) examined relationships between a broad range of indicators, or (2) examined relationships between indicators taken from the same individuals in more than one environment. We housed 60 hens in groups (48) or individually (12) for six sequential 35-day phases in different pen environments. Environments were experienced in an order counterbalanced between groups. During each phase a series of behavioural and physiological measures was taken for every bird. Indicators recorded were body and plumage condition, temperature, behaviours observed in the home pens, behaviours observed during novel object and resource competition tests, tonic immobility, physiological blood profiles, and faecal sample composition. We found that most variability in nearly all measures was not explained by either individual bird or grouping effects but varied across phases within the birds. Acknowledging this, we examined correlations between all parameters at the phase within-bird level, selecting a very conservative p-value as a cut-off point to avoid Type 1 error. A consistent set of correlations (all at least 0.24, $P<0.01$) showed that a slow approach response and alert behaviour in the novel object test was associated with higher bodyweight, lower body temperature and lower acute phase protein, heterophil:lymphocyte ratio and blood glucose level. A cluster analysis confirmed these correlations. Other important parameters known to be linked to the hens' environmental preference (eg home pen comfort behaviour) appeared to be independent of the set described above. We conclude that statistical techniques can reveal patterns of independence and redundancy in the collection of behavioural and physiological measures of coping. These result provide the basis for the development of hypothesese about welfare indicator relationships in other contexts.

Heart rate variability during allogrooming in dairy cows: a pilot study

Gutmann, Anke, Stockinger, Barbara and Winckler, Christoph, University of Natural Resources and Applied Life Sciences, Division of Livestock Sciences, Gregor-Mendel-Str. 33, 1080 Vienna, Austria; anke.gutmann@boku.ac.at

Allogrooming in dairy cows often is considered positive and relaxing. Our aim was to test this assumption by investigating effects on heart rate variability (HRV). HRV represents the current influence of the autonomic nervous system and therefore allows some insight into an animal's internal state. Our study was conducted in a herd of 20 Simmental dairy cows kept in a sloped floor system using Polar Equine s810i heart rate monitors and continuous behaviour sampling from video tapes. Heart rate (HR), root mean square of successive differences (RMSSD), and standard deviation of normal-to-normal intervals (SDNN) were calculated minute-by-minute over five minutes preceding (baseline), and the first minute during an allogrooming event (n = 76). Data were analysed using a linear mixed model including a set of factors such as role, activity, age, solicitation, and overall number of partners or events. The overall mean of baseline measures was consistent with the literature, and baseline values differed significantly depending on activity (lying (n = 48) vs. standing (n = 26): HR (LS means ± SE): 81.5 ± 2.0 vs. 95.4 ± 2.2, F = 63.3, $P < 0.001$; RMSSD: 7.0 ± .4 vs. 4.9 ± 0.4, F = 19.0, $P < 0.001$; SDNN: 14.0 ± 1.0 vs. 10.8 ± 1.0, F = 10.4, $P < 0.01$). Apart from that data showed broad variation and no consistency not only between but also within individuals, both concerning baseline and response to allogrooming. None of our predefined categories accounted for the observed differences. Changes in heart rate did not predict changes in HRV parameters and vice versa. Our results do not support a relaxing effect of allogrooming in general. Considering detailed individual and dyad data (e.g. relationship, previous encounters etc.) may be required when investigating social behaviour and its physiological effects. There is need to further investigate and accurately control for influencing and confounding factors of HRV.

Effects of environmental enrichment and loose housing of sows on behaviour and growth of weanling piglets

Bolhuis, J. Elizabeth[1], Oostindjer, Marije[1], Mendl, Mike[2], Held, Suzanne[2], Gerrits, Walter J.J.[1], Van Den Brand, Henry[1] and Kemp, Bas[1], [1]Wageningen University, Department of animal science, Marijkeweg 40, 6709 PG Wageningen, Netherlands, [2]University of Bristol, School of veterinary science, Langford House, Langford BS40 5DU, United Kingdom; liesbeth.bolhuis@wur.nl

The ability of piglets to cope with changes imposed by weaning may be affected by their rearing and housing conditions. We investigated effects of loose housing of the sow during lactation and environmental enrichment on behaviour and performance of weanling piglets. Before weaning, piglets were housed in 18 m2 enriched (E) pens with rooting materials or in 9 m2 barren (B) pens, and with a loose-housed (L) or confined (C) sow (2x2 arrangement, 32 litters in total). Piglets were mixed at weaning (day 28), with four animals per litter transferred to barren 3.2 m^2 and four to substrate-enriched 6.4 m^2 pens (n=64 groups of four piglets, 2x2x2 arrangement). Behaviour was observed at day 1, 5, 8 and 12 postweaning for 5 h per day using 2-min instantaneous scan sampling. Data were averaged per pen and analysed using GLMs. Piglets raised by loose-housed sows showed less belly nosing and other oral manipulation of pen mates than piglets from confined sows (both $P<0.05$). Piglets from enriched preweaning pens showed a higher feed intake during the first two postweaning days ($P<0.05$) and less activity ($P<0.01$) and explorative behaviour ($P<0.001$) during the 2-wk postweaning period than barren-reared piglets. Postweaning enrichment increased activity and time spent on exploratory and comfort behaviours, and reduced belly nosing and oral manipulation of pen mates (all $P<0.001$). Postweaning enrichment also increased play behaviour, with BL piglets that had switched to an enriched postweaning pen playing more than all other groups (three-way interaction, $P<0.05$). Moreover, postweaning enrichment enhanced growth ($P<0.001$), reduced number of diarrhoea days ($P<0.001$) and improved feed conversion ratio ($P<0.05$). Preweaning conditions did not affect growth or diarrhoea prevalence. In conclusion, preweaning conditions exerted enduring effects on behaviour of piglets after weaning, but particularly enrichment of the postweaning environment positively affected behaviour, health and growth of newly-weaned piglets.

The effect of environmental stressors on the social behaviour and stress physiology of group-housed growing pigs

O'Connor, Emily, Parker, Matthew, Wathes, Christopher and Abeyesinghe, Siobhan, Royal Veterinary College, Centre for animal welfare, London, AL9 7TA, United Kingdom; eoconnor@rvc.ac.uk

The physical environment of a piggery is often dim and noisy with high concentrations of atmospheric ammonia, which may interfere with the perception of visual, auditory, and olfactory cues used in social communication, thereby affecting social interactions. These conditions may represent a significant stressor to growing pigs. This study tested whether ammonia, light and/or noise affect the social behaviour and stress physiology of pigs. Four-week-old, piglets (n=224) were randomly allocated to groups of 14 and reared for 15 weeks in a 2^3 multi-factorial experiment (5 or 20ppm ammonia; 60 or 80dB(A) noise; 40 or 200lx light intensity). The frequency of aggressive (e.g., biting) and benign (e.g., joining another pig playing, feeding and/or drinking, or choosing to lie together) interactions were recorded. Play behaviour was also recorded as it is thought to reduce in stressful conditions. Production factors, e.g., growth and feed conversion ratio, were measured along with salivary cortisol and adrenal morphometry. All data were analysed using generalised linear mixed models. Compared with 5ppm ammonia, pigs exposed to 20ppm engaged in a higher percentage of aggressive interactions (14.39 ± 1.20 vs. $3.97\pm2.51\%$, $P<0.01$), spent less time playing (0.44 ± 0.16 vs. $1.20\pm0.42\%$, $P=0.03$), had lower salivary cortisol (2.95 ± 0.12 vs. 3.84 ± 0.18ng/ml, $P=0.01$) and tended to have larger adrenal cortices ($77.4\pm0.8\%$ of adrenal section vs. $75.7\pm1.4\%$, P = 0.07). Although there was no effect of dim light on stress indicators, more social interactions between pigs kept in the low light rooms were aggressive at the start of the experiment compared to the brighter rooms (14.45 ± 2.75 vs. $4.20\pm1.03\%$, P = 0.03). There was no observable impact of high noise alone on the pigs and no influence of any of the treatment conditions on production parameters. These findings suggest that ammonia, in particular, may have a negative effect on social behaviour and represent a chronic stressor.

Sow nursing behaviour in isolation and in acoustic and visual contact

Šilerová, Jitka, Špinka, Marek and Neuhauserová, Kristýna, Institute of Animal Science, Department of Ethology, Přátelství 815, Prague - Uhříněves, 104 00, Czech Republic; jsilerova@atlas.cz

Lactating sows housed in one room do not nurse independently, but rather attain high nursing synchronisation. Only few studies have examined the effect of synchronisation on nursing frequency (NF) and on occurrence of nursings without milk ejection (non-nutritive nursings, NNN). This is important practically because both NF and NNN can affect piglet milk intake. Triplets of sows with similar characteristics (farrowing date, litter size, parity) were housed in pens with the 3 treatments at the same time: Isolation (I) – visually, olfactory and auditory isolated room; Sound (S) – visually isolated pen within a room of 14 pens; Control (C) - pen with normal full contact within the room of 14 pens. Nursing behaviour (total and nutritive NF, number and proportion of NNN) of 10 triplets of sows was video recorded for 6 hours at 7 and 21 days post partum. Repeated observations on the same sow were statistically treated with a random effect. Since the acoustic communication is the decisive communicative channel for nursing synchronization we predicted that the nursing behaviour will be similar in Sound and Control treatments but different in Isolation. However sows in all 3 treatments nursed with the same total ($F_{(2, 30)}$= 1.88, P>0.1, means: I= 8.3, S=7.4, C=8.9 nursings) or nutritive ($F_{(2, 30)}$= 0.5, P>0.5, means: I=6.3, S=6.1, C=6.5 nursings) NF. Also the number ($F_{(2, 30)}$ = 1.79, p>0.1, means: I=2.1, S=1.2, C=2.3 nursings) and proportion ($F_{(2, 30)}$ = 1.79, P>0.1, means: I=24%, S=13%, C=21% of nursings) of NNN was the same. We conclude that sows housed together in one room were nursing and letting down milk at the same frequency as isolated sows and therefore the milk supply for piglets was probably similar. Detailed analysis of nursings synchronization could shade more light on this question.

Feed barrier design affects behaviour and adrenocortical activity of goats

Nordmann, Eva M.[1], Keil, Nina M.[2], Schmied, Claudia[1], Graml, Christine[1], Aschwanden, Janine[2], Palme, Rupert[3] and Waiblinger, Susanne[1], [1]Institute of Animal Husbandry and Welfare, University of Veterinary Medicine, Veterinärplatz 1, A - 1210 Vienna, Austria, [2]Federal Veterinary Office, Center for Proper Housing of Ruminants and Pigs, Agroscope Reckenholz-Tänikon ART, Research Station ART, Tänikon, CH - 8356 Ettenhausen, Switzerland, [3]Department of Biomedical Sciences/Biochemistry, University of Veterinary Medicine, Veterinärplatz 1, A - 1210 Vienna, Austria; eva.nordmann@vetmeduni.ac.at

Feed barriers for goats differ with regard to ease of entry and exit, backward view and presence of separation. The aim of our study was to investigate whether the type of feed barrier influences agonistic behaviour and stress. The study involved 55 adult, female goats of several Swiss breeds. Three groups of 14 and one group of 13 goats (2 horned, 2 hornless) were rotated between four pens with different types of feed barrier (neckrail, metal palisade, wooden palisade, diagonal fence). Each group stayed four weeks with each feed barrier type. Social interactions in the feeding area were recorded 12h per group and feed barrier (1.5h on 8 days each group) and corrected for the number of animals present in this area. Individual faecal samples were taken for analysing concentration of cortisol metabolites. Data were analysed by linear mixed-effect models taking into account interactions between feed barrier type and presence of horns. Hornless goats displayed most agonistic behaviour with body contact at the neckrail (median: 5.8 interactions/goat x 12h) and least in metal palisade (2.8, $P=0.0001$), whereas goats with horns showed far less of those interactions, thus no difference depending on feed barrier was found (from 0.7 in diagonal fence to 1.5 in wooden palisade). Agonistic behaviour leading to a loss of feeding place was displayed most at the neckrail as well (3.5, metal palisade: 2.2, $P<0.0001$). The concentration of faecal cortisol metabolites tended to be lowest for groups in the pen with metal palisade (median: 145.7ng/gr, highest in neckrail: 170.8ng/gr, $P=0.06$). In conclusion, the neckrail seems to be the least suitable feed barrier for goats, regardless if they are horned or hornless, whereas particularly the metal palisade showed the most beneficial results and therefore can be recommended. We acknowledge funding by BMLFUW and BMGF, Project Nr.100191.

Comparing the effects of different roughage diets on grooming behaviour and hairball prevalence in veal calves

Webb, Laura[1], Bokkers, Eddie[1], Heutinck, Leonie[2] and Van Reenen, Kees[2], [1]Wageningen University, Animal Science, Animal Production Systems Group, P.O. Box 338, 6700 AH Wageningen, Netherlands, [2]Wageningen University and Research Centre, Livestock Research, P.O. Box 65, 8200 AB Lelystad, Netherlands; laura.webb@wur.nl

Veal calves fed roughage seem less likely to develop hairballs in their rumens compared to calves fed all-liquid diets. Hairballs may impair digestion, hence reduce welfare. This study examined the effects of different diets on grooming behaviour and hairball prevalence, investigating whether grooming time could explain hairball incidence. In a 3x2x2 factorial design, 240 group-housed calves were fed three roughage types (maize silage, maize cob silage or straw), in two amounts (250 or 500g DM/day) and two particle sizes (chopped or ground), supplemented to milk replacer (MR) from 4 weeks of age. Control treatments (N=20/treatment) were MR-only, MR+iron and MR+ad libitum hay. Grooming behaviour of self and other calves was obtained from video recordings (11h/day) using scan sampling at 19 and 24 weeks. Calves were slaughtered at 25 weeks and the number of animals with ruminal hairballs was recorded. The factors type, amount and particle size had no significant effect on grooming behaviour. At 19 weeks, MR-only calves spent more time grooming than calves fed straw, maize silage or hay (Mean%scans±s.e.m.: 14±1.1 versus 10±0.5, 11±0.6 and 8±0.9 respectively; GLM, $P<0.05$). Compared with maize cob silage, ad libitum provision of hay reduced grooming time (8±0.9 versus 12±0.6; GLM, $P<0.05$). No significant effect of diet was found on grooming activity at 24 weeks. Significantly less calves fed roughage had ruminal hairballs at slaughter compared with calves fed all-liquid diets (straw and hay: 0%, maize silage: 14%, maize cob silage: 30% versus MR-only and MR+iron: 85%; logistic regression: $P<0.001$). Significantly less calves had hairballs in groups fed chopped compared to ground roughage (9% versus 20%; logistic regression: $P<0.05$). Differences in grooming behaviour between treatments were not systematically associated with differences in hairball prevalence. Present results support the idea that roughage, particularly longer particles, increases rumen motility thereby promoting continuous removal of ingested hair.

Performance of low and normal birth weight piglets in a cognitive holeboard task

Gieling, Elise, Park, Soon Young, Nordquist, Rebecca and Van Der Staay, Franz Josef, Utrecht University, Faculty of Veterinary Medicine, Farm Animal Health, Yalelaan 7, 3584 CL Utrecht, Netherlands; e.t.gieling@uu.nl

We questioned whether being born with a low birth weight would negatively affect cognitive performance of piglets in a spatial holeboard task. To examine this, nine pairs of siblings (♀) were tested starting at the age of 9 weeks. The pairs consisted of a low and normal birth weight animal. Before testing the animals were trained until they were confident in the apparatus. Rewards were hidden in food bowls covered by plastic balls which the pigs could lift upwards with their snouts. Each pig was tested in its own individual configuration of 4 out of 16 baited holes. For the 1st configuration of baited holes, animals were tested twice a day during 13 weekdays (spaced trials). Next, they were trained on two reversals in which food was hidden in new configurations. Results clearly show that both groups are very well able to learn the configurations of baited holes. All animals rapidly reduced the number of visits to unbaited holes (reference memory) and of revisits to the baited set of holes (working memory). The animals with a normal birth weight acquired the WM component of the first reversal faster than their low birth weight siblings after the first reversal ($P<0.0117$). This effect disappeared with further training. In commercial pig reproduction systems, post-natal mortality is common. However, a considerable number of light weight piglets survive the critical period after birth. The long-term implications of being born light or underweight in pigs are unknown, but from human studies we know that for example learning is impaired. There is two aims studying low birth weight in relation to cognitive deficits in pigs; the species could be used as a model animal for cognitive deficits in humans and on the other hand, gaining more knowledge about pig cognition in general will benefit animal welfare.

Dopamine regulation of reward-related behaviours in laying hens

Moe, Randi O.[1], Nordgreen, Janicke[1], Janczak, Andrew[1], Zanella, Adroaldo J.[1,2] and Bakken, Morten[2], [1]Norwegian School of Veterinary Science, Department of Production Animal Clinical Sciences, P.O. Box 8146 dep., 0033 Oslo, Norway, [2]The Norwegian University of Life Sciences, Dept. of Animal- and Aquacultural Sciences, P.O Box 5002, 1432 Ås, Norway; randi.moe@nvh.no

The experience of positive affective states is important for good animal welfare, and indicators of such affective states are needed. Behaviours expressed during anticipation and consumption of a signalled positive reward are candidate indicators as they are thought to be regulated by dopaminergic and opioid activation in the brain reward system, respectively. The neurochemical regulation of anticipation has not yet been tested in poultry. In the present study we tested the hypothesis that dopamine regulates anticipatory, but not consummatory behaviour in a classical conditioning paradigm in laying hens. Hens (N=10) were trained to associate a light signal (CS) with a meal worm reward (US). We tested the effects of saline and four different doses of haloperidol (0.3-2.0 mg/kg), a dopamine D2 receptor antagonist, injected 30 min before CS, on locomotor, anticipatory and consummatory behaviours during the CS-US interval and after reward delivery. Treatment effects were tested using a mixed model ANOVA (behaviour = treatment + repetition + hen identity) with hen identity as a random effect. Head movements were decreased in all treatments compared with controls ($P<0.0004$). There was also a significant effect of treatment on latency to start anticipatory behaviour ($P<0.03$), with a significantly shorter latency in the saline group compared with the treatments 3 and 5. Haloperidol treatment tended to decrease duration of anticipatory behaviour ($P<0.06$). Locomotion was unaffected by haloperidol even at the highest doses expected to induce sedation (ns). This highest dose of haloperidol increased the latency to peck at the reward compared to all other doses ($P<0.04$), indicating some effect of sedation. Thus, dopamine may have a role in the regulation of reward-related behaviour in laying hens, indicating that anticipatory behaviour as defined in this study may be a candidate indicator of positive affective states in poultry.

Object habituation in horses

Christensen, Janne Winther, Aarhus University, Animal Health and Bioscience, Blichers allé 20, 8830 Tjele, Denmark; jannewinther.christensen@agrsci.dk

The ability of horses to habituate to various stimuli greatly affects safety in the horse-human relationship. Object habituation appears to be rather stimulus specific, however, and shape and colour have been shown to be important factors for object generalisation. In this experiment, we investigated whether test horses (n=15) that were habituated to a complex object (composed of five objects of varying shape and colour) subsequently reacted less to (1) objects that were previously part of the complex object and (2) a novel object (representing a new shape and colour), compared to CONTROL horses (n=15). The 30, 2-years-old Danish Warmblood mares were habituated to social separation in the test arena prior to the experiment, and were randomly assigned to the treatment groups. The horses were tested individually and feed containers were used to encourage the horses to approach the objects. Behavioural reactions towards the objects, latency to eat and heart rate (HR) of test and control horses were analysed in a one-way ANOVA (SigmaStat 3.1). Compared to controls, test horses reacted significantly less towards objects, which were previously part of the complex object (e.g. mean HR (bpm); test: 60.3 ±2.0 vs. control 76.5 ±4.7, F=10.6, P=0.003), indicating object recognition. In contrast to our expectations, test horses also reacted significantly less towards the novel object of new shape and colour (e.g. mean HR; test: 62.8 ±2.6 vs. control 80.6 ±5.4, F=9.1, P=0.006), suggesting that test horses were capable of object generalisation. In conclusion, test horses demonstrated object recognition and further generalised the habituation to a novel object of new shape and colour. Thus, it appears possible to increase object generalisation in horses by habituating them to a range of colours and shapes simultaneously. This knowledge greatly affects how horses may be trained to react calmly towards frightening objects.

How do pigs react to situations of different emotional valence?

Mueller, Sabrina[1], Van Wezemael, Lea[1], Gygax, Lorenz[2], Stauffacher, Markus[1] and Hillmann, Edna[1], [1]Institute of Plant, Animal and Agroecosystem Siences, ETH Zurich, Animal Behaviour, Health and Welfare, Universitaetsstrasse 2, CH 8092 Zurich, Switzerland, [2]Federal Veterinary Office, Centre for Propper Housing and Pigs, Agroscope Reckenholz-Taenikon, Research Station ART, Taenikon, CH 8356 Ettenhausen, Switzerland; sabrina-mueller@ethz.ch

Positive anticipation may increase welfare in pigs. Thus, indicators reflecting emotional reactions during anticipation are needed. The aim of this study was to examine whether such indicators differ during anticipation of either unpleasant or pleasant situations. Sixteen weaned pigs were tested individually in both, a postulated positive and negative situation. After 60 s of waiting, one of two tone sequences was played, followed by an anticipation period of 10 s before a door to an experimental box was opened. Depending on the tone sequence the pig either had access to rewarding popcorn (positive situation) or had to cross a wooden ramp without reward (negative situation). At first, they were conditioned to a tone sequence corresponding to the positive situation 14 times and eight times to the negative tone sequence. Anticipation was then increased stepwise to 10 s. Vocalisation, locomotion and different heart beat parameters were analysed before and during the tone sequence was played, during anticipation and at the end of each trial. Data were analysed using linear mixed effects models. The proportion of pigs uttering high-frequency vocalisation was higher during anticipation of the unpleasant situation ($F_{2,206}$=13.1; $P<0.0001$). After the door was opened, pigs hesitated longer before moving ($F_{1,13}$=68.1; $P<0.0001$) and turned around more often (t_{13}=9.8; $P<0.0001$) in the unpleasant situation. No differences were found in the heart beat parameters. The pigs showed different behavioural responses during anticipation and experience of two situations of different emotional valences as expected. However, in the heart beat parameters no difference was found during anticipation. The reasons for this could be a too short training period or difficulties in learning the association between an acoustic signal and a corresponding situation. On the other hand, both situations could have induced arousal which was not discriminable according to the used measures.

Will work for space: broiler chickens are motivated to move to lower stocking densities

Buijs, Stephanie[1,2], Keeling, Linda J.[2] and Tuyttens, Frank A.M.[1], [1]Institute for Agricultural and Fisheries Research (ILVO), Animal Sciences Unit - Animal Husbandry and Welfare, Scheldeweg 68, 9090 Melle, Belgium, [2]Swedish University of Agricultural Sciences, Department of Animal Environment and Health, P.O. Box 7068, 75007 Uppsala, Sweden; stephanie.buijs@ilvo.vlaanderen.be

There is continuing debate concerning suitable space allocations for broilers, but this discussion seldom includes the birds' own preferences. A new methodology is presented that quantifies preferences by studying broilers' choices between different stocking densities. Densities were separated by a barrier to deter non-motivated birds from moving between densities. Two barrier heights were used, which were determined in a parallel experiment in which broilers crossed barriers to feed. The maximum height that ≥75% of the tested individuals crossed to feed was used as a 'low barrier' treatment in the density experiment. By repeating the feeding experiment after 6 hours of feed deprivation, a 'high barrier' treatment for the density experiment was determined. Barrier heights were determined weekly. Density preference was tested during weeks 4 to 6 of life, using four groups of 104 birds each. Birds were housed in pens consisting of two compartments. The shape of one these compartments could be adjusted, thus creating different space allocations/densities. For each test, half of the group was moved to the adjustable compartment (densities: 9.3, 12.1 or 14.7 birds/m^2), and the other half to the fixed compartment (14.7 birds/m^2). After habituation, movement between the compartments was scored. More birds moved into the adjustable compartment than out of it, even in treatments using the 'high barrier', which had deterred 20 to 25% of feed-deprived birds from feeding. The relative movement into the adjustable compartment (birds moving in - birds moving out) increased with size of the adjustable compartment, and this effect increased with age (LSMEANS for movement to the lowest, medium and highest densities in week 4: 11, 7, 4; week 5: 13, 8, 4: week 6: 14, 6, 1, $F_{4,131}$=3.01, P=0.02). Thus, broilers showed considerable motivation for densities below 14.7 birds/m^2, which was already slightly lower the maximum of the EU Directive (42kg/m^2).

A happy hen? Comparing two different measures of emotional state

Wichman, Anette[1], Keeling, Linda J.[2] and Forkman, Björn[1], [1]University of Copenhagen, Department of Large Animal Sciences, Grönnegårdsvej 8, 1870 Frederiksberg C, Denmark, [2]Swedish University of Agricultural Sciences, Animal Environment and Health, P.O. Box 7068, 750 07 Uppsala, Sweden; anette.wichman@hmh.slu.se

The performances of 31 adult laying hens in a cognitive bias and an anticipation test, both designed to measure the assumed emotional state of the birds, were investigated with the aim of comparing the tests. Initially birds were group housed in 10 pens (2.55m^2) with nest box, perches, litter, food and water. Five days before the first test period half the pens were converted to enriched housing (received straw, hay, peat, higher perches and extra food) and half to poor housing (received less litter and a lower perch). In the first test period half the birds were tested in the cognitive bias test and half in the anticipatory test and in the second test period, carried out seven weeks later, this was reversed. The reward was corn in both tests. In the cognitive bias test the birds were trained that a given bowl to the left (or right - balanced by bird) contained corn and latency to approach three ambiguous probes, placed at distances between the rewarded and unrewarded locations, was measured. In the anticipation test birds were trained that a light cue signalled a 25 second delay to the reward and behaviour was observed during the delay to investigate if enriched birds displayed a decreased sensitivity to the reward. No difference between treatments was found for the first period for either test. In the second period there was no difference between treatments in the anticipation test, but birds in the enriched environment were slower to approach the near-rewarded probe in the cognitive bias test (Mann-Whitney test; P=0.015). The results suggest that the chosen treatments did not induce large differences in the birds´ assumed emotional state, but that the cognitive bias test was perhaps the more sensitive test to pick up those differences that did occur over time.

Comparing the influence of pre and post-consumption phases of a rewarding event on cognitive bias in dogs

Burman, Oliver[1,2], Trudelle Schwarz Mcgowan, Ragen[1], Mendl, Michael[2], Norling, Yezica[1], Paul, Elizabeth[1,2], Rehn, Therese[1] and Keeling, Linda J.[1], [1]Swedish University of Agricultural Sciences, Department of Animal Environment and Health, Uppsala, Sweden, [2]University of Bristol, School of veterinary science, Langford House, Bristol BS40 5DU, United Kingdom; oburman@lincoln.ac.uk

As interest in the positive emotional states of non-human animals increases, it is timely to carry out research into the different types of positive emotional states that animals might experience and how/when they are generated. We used a cognitive bias judgement test to investigate whether a low arousal positive emotional state (e.g. satisfaction/contentment) could be induced in domestic dogs as a result of their experiencing a food-based foraging task. In this task, subjects (1yr old female Beagles) had to search for small amounts of food randomly placed within a maze arena. We predicted that dogs tested during the post-consumption phase of this rewarding event (i.e. soon after experiencing the foraging task), should be more likely to judge ambiguous stimuli positively compared to when tested during the pre-consumption phase without going on to experience the task. Using a balanced within-subjects design, the dogs (N=12) received the cognitive bias test either without experiencing the foraging task (Pre-consumption), or immediately after experiencing the foraging task (Post-consumption). When analysing the dogs' response to the ambiguous (probe) stimuli, a repeated measures GLM revealed a strong trend towards a Treatment x Probe interaction ($F_{2,10}=3.6$, $P=0.065$), but no overall Treatment effect ($F_{1,11}=0.2$, $P=0.7$). Further investigation of the Treatment x Probe trend revealed that dogs took longer to approach the middle ambiguous probe during Post-consumption, compared to during Pre-consumption (Post-consumption: 7.9±1.3secs; Pre-consumption: 5.2±0.83secs. t11=-2.66, $P=0.022$). This result suggests that, contrary to our prediction, dogs showed a more negative judgement during Post-consumption, perhaps due to a reduced motivation for an uncertain food reward. Alternatively, the dogs may have become excited during Pre-consumption due to anticipating the foraging task. This could be investigated in future studies with controls that allow for differences in arousal. Further work is therefore necessary to clarify the influence of the reward cycle on positive emotional state induction.

A measure of sensory (visual) lateralization in dogs (*Canis familiaris*): the Sensory Jump Test

Tomkins, Lisa[1], Williams, Kent[2], Thomson, Peter[1] and McGreevy, Paul[1], [1]University of Sydney, Faculty of Veterinary Science, R.M.C. Gunn Building B19, University of Sydney. NSW, 2006, Australia, [2]SCUZ Technologies, Wetherill Park, NSW, 2164, Australia;
tomkins@optushome.com.au

Asymmetries of motor function have been widely reported in a variety of species, along with associations between behavioral tendencies and lateralization. Right-preferent animals are described as more willing than left-preferent animals to approach unfamiliar stimuli. Only limited literature is available on lateralization of sensory functions, including visual laterality. The current study is one of the first to investigate this attribute in dogs. Sensory lateralization was assessed in dogs (n = 74) using our innovation, the Sensory Jump Test (SJT). Modified head halters imposed three ocular treatments (binocular vision [BV], right monocular vision [RMV], and left monocular vision [LMV] to assess eye preference in the SJT. Dogs completed 10 jumps (a jump set) for each treatment, and each dog was subjected to all three treatments in a randomized sequence. Measurements included (1) launch and landing paws, (2) type of jump, (3) approach distance [AD], (4) clearance height of forepaw [CHF], hindpaw [CHH], and lowest part of body [LPB] to clear jump, and (5) whether the jump was successful. Several factors including ocular treatment, jump set and replication number were significantly associated with jump outcomes. In the first jump set, findings indicated a left hemispheric dominance. Jump kinematics were compromised in dogs confined to LMV compared with those confined to RMV and BV, with a significantly reduced AD and CHF observed in dogs with LMV. However, by the third jump set, compared with RMV and BV dogs, LMV dogs launched further back and had higher CHH and LPB, corresponding to an increase in jump success. Our measures of sensory (eye preference) and motor laterality (paws used in jumping) showed no association within individuals. This novel report on canine sensory lateralization is the first to show that it is organized on a different level to motor laterality in the brain of dogs.

Calling individual sows to feed: a possibility to reduce agonistic interactions in electronic sow feeding systems?

Kirchner, Jasmin[1], Manteuffel, Gerhard[2] and Schrader, Lars[1], [1]Friedrich-Loeffler-Institut, Institute of Animal Welfare and Animal Husbandry, Doernbergstrasse 25/27, 29223 Celle, Germany, [2]Leibnitz Institute for Farm Animal Biology, Behavioural Physiology, Wilhelm-Stahl-Allee 2, 18196 Dummerstorf, Germany; jasmin.dannenbrink@fli.bund.de

An electric sow feeding system (EFS) enables control over individual feed intake and facilitates the management of group housed gestating sows. In particular in dynamic groups are often problems with agonistic interactions around the feeding station when different sows try to enter. In this study we conditioned gestating sows for different play backed names to call them individually to the feeding station and hereby, reducing agonistic interactions. A dynamic group of 60 sows of different lactations was kept in a three week production rhythm with seven subgroups. In two trials lasting 30 weeks each the gestating sows were housed in a pen with littered lying areas, an outdoor run and EFS. The first trial with about 35 naïve sows served as a control. In the second trial the EFS was modified to a call feeding station (CFS). Before this trial all sows were trained in subgroups of 6 to 8 sows in a separate smaller pen and then kept in the large pen (32 sows) where they were called individually once a day in a random order by the CFS. In both trials individual agonistic behaviours were observed by video over a period of eight days in an area of 13 m^2 at the entrance of the feeding station. Preliminary results indicate that the sows were able to learn their names within one week and that they remember their calls six weeks after leaving the gestation pen. Agonistic interactions were lower when the sows were fed with the call feeding station (0.15±0.09 per hour per sow) compared to the control (0.65±0.29) ($P<0.01$). In conclusion, a call feeding station can reduce agonistic interactions even in larger groups of gestating sows and may increase the sows' welfare by cognitive enrichment.

Provision of designated defecation areas to reduce the spatial distribution of faeces by grazing swamp buffalo heifers

Somparn, P.[1], Juengprasobchok, P.[1], Faree, S.[2] and Sawanon, S.[3], [1]Thammasat University, Deaprtment of Agricultural Technology, Faculty of Science & Technology, 99 Paholyothin road, Amphur Klong Luang, Pathumthani, 12121, Thailand, [2]Department of Livestock Development, Surin Livestock Research and Breeding Center, Tambon Na-Bua, Amphur Muaeng, Surin, 32000, Thailand, [3]Kasetsart University, Department of Animal Science, Faculty of Agriculture, Amphur Kamphaengsaen, Nakhon Pathom, 73140, Thailand; somparn@tu.ac.th

Domesticated buffaloes stocked within small paddocks usually defecate over the whole area. The aim of this experiment was to examine whether the total area over which faeces are distributed in a small paddock can be reduced by offering a 'designated dung area' for defecation. Four groups of three 2-year-old swamp buffalo heifers were each subjected to four treatments, during four 7-day periods in a 4×4 Latin square design experiment, replicated in time. The treatments provided a pasture which included either a designated dung area and wallow (DW), only a designated dung area (D), only a wallow (W), or neither a designated dung area nor a wallow (N). The designated dung areas measured 2 m × 2 m and were created by spreading approximately one cubic meter of faeces, which had been collected before the start of experiment, over the area. Buffalo behaviour was scan sampled at 1-min intervals over 24 hours (commencing 06:00 h, day 3) during each period. All defecations were surveyed and mapped using the ArcView program. The results show that there was no treatment effect on time spent grazing, ruminating or idling ($P>0.05$). Buffaloes averaged 5 defecations per day, with a mean weight of 0.27 kg dry matter and covering a mean area of 0.14 m^2. Providing a wallow (treatments DW and W) had little effect of concentrating defecations within a smaller area, with less than 20% of defecations occurring within a 10-metre radius of the wallow. In contrast, when a designated dung area was provided (treatments DW and D) more than 80% of all defecations occurred within a 10-metre radius of that area. On treatment N, defecations were widely spread across the paddock. Results show that providing a designated dung area can reduce the area over which faeces are distributed by grazing swamp buffaloes.

Short-term overcrowding affects the behavior of lactating Holstein dairy cows negatively

Krawczel, Peter D.[1,2], Klaiber, Laura B.[1], Butzler, Rachel E.[1], Klaiber, Lisa M.[1], Dann, Heather M.[1] and Grant, Richard J.[1], [1]William H. Miner Agricultural Research Institute, 1034 Miner Farm Road, P.O. Box 90, Chazy, NY 12921, USA, [2]The Univesity of Vermont, Department of Animal Science, 102 Terrill, 570 Main, Burlington, VT 05405-0148, USA; krawczel@whminer.com

The objective of this study was to determine the effect of stocking densities of 100 (1 cow per freestall and headlock), 113, 131, and 142% on the behavior of Holstein dairy cows. Multiparous (n = 96) and primiparous cows (n = 40) were assigned to 4 pens (n = 34) and treatments were imposed for 14 d using a 4 × 4 Latin square design. Feeding and ruminating were quantified by 24 h of direct observation of focal cows (n = 12 per pen) at 10-min intervals on d 11. Focal cows were selected to represent pen means of days in milk, milk production, and parity and the pen range of body weighs. Dataloggers recorded lying behavior of focal cows at 1-min intervals on d 10 to 14. Displacements from the feed bunk were recorded for 2-h intervals (n = 9) continuously after milking on d 11 to 14 at the pen-level. Data were analyzed using the MIXED and REG (variables with significant F-tests) procedures of SAS. Feeding (3.7 ± 0.2 h/d), ruminating (7.2 ± 0.2 h/d), and lying bouts (11.3 ± 0.6 per d) did not differ among treatments ($P>0.5$). Of the total observed rumination, the percentage occurring within freestalls decreased linearly (y = -0.19x +114.6; R^2 = 0.27; $P=0.02$) from 95.1 ± 2.4% at 100% (stocking density) to 87.3 ± 2.4% at 142%. Lying time decreased linearly (y = -0.02x + 14.7; R^2 = 0.26; $P=0.02$) from 12.9 ± 0.2 h/d at 100% to 12.3 ± 0.2 h/d at 142%. Displacements increased linearly (y = 0.27x – 18.5; R^2 = 0.60; $P<0.01$) from 7.5 ± 0.9 per 2 h at 100% to 20.5 ± 1.8 per 2 h at 142%. The observed behavioral changes suggest short-term overcrowding alters the time budgets of lactating dairy cows negatively. The long-term effects of these changes are unknown and should be evaluated to fully understand the relationship between overcrowding and welfare.

Domestication effects on behavioural synchronization in chickens

Eklund, Beatrix and Jensen, Per, Linköping Univeristy, Biology, Linköping University, SE-581 83 Linköping, Sweden; beaek@ifm.liu.se

During domestication, animals have adapted to a protected environment. Behavioural synchrony (allelomimetic behaviour) is one aspect of anti-predator strategies which may have been affected. Chickens are known to adjust synchronization and inter-individual distances depending on behaviour. We hypothesized that both aspects have changed during domestication, and that White Leghorn (WL) chickens would show less synchronized behaviour and less distance sensitivity than the ancestor, the red jungle fowl (RJF). In this study 60 birds, 15 female and 15 male WL and the same number of RJF (28 weeks old) were studied in groups of three in pens (1m × 2m) for 24 consecutive hours, following 24 hours of habituation. The pens contained feed, water, and a perch, and the floor was covered with wood chips. Video tapes covering four hours per group (dawn, 9-10 a.m, 1-2 p.m and dusk) were analysed, and we recorded the frequency of different behaviours, the degree of synchrony, and the individual distances taking account of the bigger body size in WL. There were no breed effects on the frequency of different activities, and no effect on the overall frequency of behaviour where two or three birds were synchronized. Neither was the synchrony of behaviour or the average distance between the chickens affected by breed. After performance of the two most synchronized behaviours, individual distance increased more for RJF than WL($F(1,13)=4.8$, $P=0.047$ for perching; $F(1,13)=4.7$, $P=0.046$ for comfort behaviour). According to this study domestication appears not to have significantly altered the relative frequencies of different activities, or the degree of behavioural synchronization in chickens. However, the results indicate that WL may be less sensitive to the individual distanxce when performing certain behaviours, possibly as an adaptation to domestication.

Flocking for food or flockmates? A model of chicken social behaviour

Asher, Lucy[1], Collins, Lisa M.[1], Pfeiffer, Dirk U.[1] and Nicol, Christine J.[2], [1]Royal Veterinary College, VCS, Royal Veterinary College, Hawkshead Lane, North Mymms, Hatfield, AL9 7TA, United Kingdom, [2]University of Bristol, Department of Clinical Veterinary Sciences, Langford House, Langford, BS40 5DU, United Kingdom; lasher@rvc.ac.uk

Animals in groups behave in a cohesive manner. But how is group cohesion maintained? Do animals move in an assertive manner, according to their own motivations, or in a social manner, with respect to the movements of others? We used an agent-based mathematical model to determine the contributions of social and assertive factors in determining the movements of groups of four laying hens housed in small pens. To distinguish between these factors, the behaviour of agents in the model was determined by a social weighting factor which could range from 0-1 and deviation in resource attraction between agents. Leader-follower systems were modelled such that one agent was highly assertive (low social weighting factors) and the other agents were highly social (high social weighting factors). The fit of the model was determined through comparison with observed clustering of hens, using bootstrapped wilcoxon tests where higher P values indicate better model fit. Clustering was measured by the maximum number of hens/agents in any of six areas of the pen, divided by the total number of areas occupied. The results of the model suggest that group cohesion in laying hens is determined primarily through resource attraction. We found the best model fit when the social weighting factor of all agents was low at either 0 ($P= 0.940$) or 0.1 ($P= 0.789$). Model fit to observed data was low ($P<0.05$) when the model had leader and follower agents. Therefore, there was no evidence for leaders and followers in the hens we observed. These results suggest that in the hens we observed, hens were not moving in a truly socially cohesive way, with respect to the movements of others, but rather were exhibiting clustered behaviour due to shared resource attraction.

Selection for reduced fear of humans in red junglefowl: correlated effects on behaviour and production traits

Jöngren, Markus and Jensen, Per, Linköping University, IFM Biology, Division of Zoology, Linköping University, 58183 Linköping, Sweden; markj@ifm.liu.se

It is believed that one main driving force causing domestication is selection for a reduced fear response towards humans. Earlier studies indicate that the so called domesticated phenotype may occur as a correlated cascade of responses to this selection. In the present study we aim to uncover the genetic mechanisms behind domestication using the red jungle fowl as a model. Starting from an outbred population, a standardized fear-of-human-test was used to categorize the animals according to high, intermediate or low fear response. The high and low response birds were selected for breeding, resulting in two lines of birds with high or low fear response towards humans, whilst the intermediate group was maintained without further selection. The plans are that selection will be done in each of five generations of 150 animals each at the same time as all animals are put through a series of behavioural tests examining correlated responses in sociality, aggressiveness, explorativity, feeding behaviour and fear responses. So far, correlations between traits have been investigated in the parental generation. The first generation yielded a high and low line that were significantly different in the fear-of-human (selection) index (2,41 vs. 1,56 out of 3; $P<0.001$; t-test). Further the selection index correlated with rightening time in a TI test($r=0,43$; $P<0.05$; Pearson correlation). Also correlations were found between freezing behaviour in an Aerial predator exposure test and egg production ($r=0,35$; $P<0.05$) and between hatch weight and explorativity in an open field test ($r=0,36$, $P<0.05$). The results indicate that fear behaviour generalises across different tests, and may be related to production traits. After sacrifice, gene expression analysis by means of microarray will be preformed on selected parts of the brains, providing a longitudinal comparison between changes in behaviour and gene expression over generations.

Do hens have friends?

Abeyesinghe, Siobhan[1], Asher, Lucy[2], Collins, Lisa[1], Drewe, Julian[1], Owen, Rachael[1] and Wathes, Christopher[1], [1]Royal Veterinary College, Department of Veterinary Clinical Sciences, Hawkshead Lane, North Mymms, Hatfield, AL9 7TA, United Kingdom, [2]University of Nottingham, School of Veterinary Medicine and Science, Sutton Bonington Campus, Loughborough, LE12 5RD, United Kingdom; sabeyesinghe@rvc.ac.uk

Recent emphasis on positive welfare has generated interest in the capacity of farmed animals to form socio-positive bonds, which might enrich their lives and afford social support under stress. Social behaviour of eight identical pens of 15 hens (aged 18 weeks at commencement) was recorded over eight weeks. Pens (1.75m × 1.75m × 2m high) were bedded with wood-shavings and offered ad libitum water, grit and commercial pelleted food, a 4-hole front roll out nest box and 2.4m of perch space. Two temporally independent data sets were used to investigate dyadic relationships characterised by consistent geographical and temporal associations in different contexts. Social network analysis was performed on adjacency matrices for both existence (binary) and number (weight) of individual proximities within four resource-based areas, generated from 5 min scan-samples across two 15 min periods (immediately and 6 h after lights on) per pen per week. No evidence was found for stability in association networks over time, suggesting that hens did not preferentially locate with others in the defined resource areas. Though significant (Bonferoni corrected) correlations were seen in existence data across adjacent pairs of weeks for some pens (e.g. pen 1, weeks 3 and 4, r=0.297, P=0.006), suggesting that individual birds associated with particular individuals for a short period, these associations did not persist across longer time-frames. Final perch-roosting positions, recorded at dusk three nights per week for the same eight weeks, were compared with data generated from a sample-without-replacement model. Paired roostings occurred less often than expected for perching at random (mean±SD observations recorded in proximity: model 0.12±0.01 vs. observed range 0.01±0.03 to 0.04±0.04, P<0.001), but could be explained by not all birds being recorded on the perch. In general there were no positional preferences in any pen (left vs. right; 2-tailed t-tests P>0.05). In conclusion, no evidence was found for consistent dyadic associations between pairs of individual hens. Further work will examine whether this is due to conspecific indifference or uniform sociality.

Perch material and diameter affects perching behaviour in laying hens

Pickel, Thorsten[1,2], Scholz, Britta[1] and Schrader, Lars[1], [1]Friedrich-Loeffler-Institut, Institute of Animal Welfare and Animal Husbandry, Doernbergstrasse 25/27, 29223 Celle, Germany, [2]University of Muenster, Department of Behavioural Biology, Badestrasse 9, 48149 Muenster, Germany; britta.scholz@fli.bund.de

Perching is an important behaviour for laying hens. However, the provision of perches is associated with health problems which may result from inadequate perch designs. This study focuses on particular behavioural patterns shown by laying hens on different perches during night-time in order to test whether perching behaviours may indicate the suitability of a perch design. 60 Lohmann Selected Leghorn hens were cage-reared without perches and kept from 18 to 29 weeks of age in six compartments over two identical trials. Hens were randomly offered nine circular perches, which differed in material (wood, steel, rubber cover) and diameter (27, 34, 45 mm) for one week each. Duration of standing, resting position (head tucked backwards into feathers or head forward with neck pulled back), preening and frequencies of balance movements and comfort behaviours shown by individual hens on each perch were video-observed throughout one night at the end of each observation week. Data were analysed using a mixed model ANOVA. Balance movements decreased with increasing perch diameter and were less on rubber perches (1.72 ± 0.33 per hour per hen) compared to wood (2.39 ± 0.33) and steel (2.30 ± 0.34) ($P<0.05$). On steel perches, hens rested longer with their heads tucked backwards into their feathers (steel: $53.1\pm2.5\%$, wood: $29.1\pm2.4\%$, rubber: $34.4\pm2.4\%$, $P<0.001$) and less with their heads forward compared to wood and rubber (steel: $31.5\pm2.3\%$, wood: $51.9\pm2.2\%$, rubber: $47.4\pm2.2\%$, $P<0.001$), irrespective from perch diameter. Standing was shown less often on steel compared to wood or rubber perches (steel: $10.2\pm1.1\%$, wood: $14.9\pm1.0\%$, rubber: $14.8\pm1.0\%$, $P<0.001$). Balance movements may be the most sensitive indicator with respect to perching stability. As steel is of higher thermal conductivity compared to wood or rubber, less standing and more resting with head tucked backwards may reflect thermoregulative behaviour in order to minimise heat loss rather than indicating the suitability of a perch.

Effects of early rearing environment on learning ability and behavior of laying hens

Croney, Candace[1], Morris, Hannah[2] and Newberry, Ruth[3], [1]The Ohio State University, Department of Veterinary Preventive Medicine, 1920 Coffey Road, 43210, USA, [2]Oregon State University, Department of Animal Sciences, 112 Withycombe Hall, Corvallis, OR 97331, USA, [3]Washington State University, Center for the Study of Animal Well-being, P.O. BOX 646520, Pullman WA 99164, USA; croney.1@osu.edu

The effects of enriching laying hens' rearing environments on their learning ability and behavior were investigated. One hundred and seventeen day-old ISA brown layer chicks were housed in groups of 20 and randomly assigned to enriched or unenriched floor pens. Enriched rooms provided decorative streamers suspended from the ceiling, classical music, nutritional enrichment (mealworms, plants, and hay), and daily human contact. Human approach tests were conducted at 2, 4, 6, 8, 10 and 11 weeks of age. At week 13, a foraging test was conducted to investigate the birds' spatial navigation abilities. Bird weights, feather scores, egg numbers, weights, and locations were recorded for 13 weeks following onset of lay at 17 weeks of age. Exact Wilcoxon rank sum tests were used to analyze the approach and foraging tests, while chi square tests were used to analyze feather score data. Enriched birds showed a tendency to be less fearful of familiar and novel humans ($P<0.10$), and foraged more efficiently, locating more food patches and demonstrating more vertical investigations than the unenriched birds ($P<0.05$). No differences were found between enriched and unenriched birds in regards to egg numbers, egg weights, or the number of eggs laid on the floor. Enriched birds broke more eggs than unenriched birds ($P<0.05$), but they also utilized their nest boxes more fully, laying more eggs in the top tiers than unenriched birds ($P<0.10$). Additionally, enriched hens weighed more than the unenriched hens ($P<0.05$) and had better feather condition ($P<0.001$). Overall, the results suggest that enrichment improved some aspects of hen behavior and performance, such as reduced fear of humans and feather pecking, and increased foraging efficiency and ability to locate and use resources in the vertical plane.

The effect of birth weight and age on the behavioural responses of piglets to tail docking and ear notching

Bovey, Kristi[1,2], Devillers, Nicolas[3], Lessard, Martin[3], Farmer, Chantal[3], Dewey, Cate[1], Widowski, Tina[1] and Torrey, Stephanie[1,2], [1]University of Guelph, Guelph, ON N1G 2W1, Canada, [2]Agriculture and Agri-Food Canada, Guelph, ON N1G 2W1, Canada, [3]Agriculture and Agri-Food Canada, Sherbrooke, QC J1M 1Z3, Canada; kbovey@uoguelph.ca

Selection for high prolificacy has resulted in litters comprised of a greater number of low-birth-weight (LBW) piglets of 1 kg or less. Given their presence in over three-quarters of litters, and increased mortality rate, it is clear that a greater understanding of LBW piglet management is required for both animal welfare and productivity. The objective of this experiment was to compare the effects of processing low-birth-weight piglets during the first 24 h versus at 3 d of age on pain-related behaviour and growth. Six piglets per litter from 20 litters (n=120 piglets) were used in a 2×2 complete block design and analyzed using mixed model procedure in SAS with sow as random effect. Piglets were weighed at birth and designated as low-birth-weight (0.6 - 1.0 kg; LBW) or average-birth-weight ($\geq$1.2 kg; ABW), and processed (tail docked, ear notched) at either 1- or 3-days of age. Vocalizations were recorded during the procedures and analyzed using Raven Pro 1.3. Average daily gains (ADG) were determined through 21 days of age. During the procedures, LBW piglets produced fewer calls/second (0.95±0.08 vs. 1.16±0.06 Hz, P=0.0147), with a lower mean frequency of high frequency calls (2,504.30±267.94 vs. 2,685.46 ±190.08 Hz, P=0.0587). Though ADG differed between LBW and ABW treatments (175.8±0.008 vs. 235.3±0.006 g/d, P<0.0001), the ADG between birth and d 5 was not affected by day of processing (200.7 ±7.0 vs. 210.4±7.0 g/d, P=0.1141). Similar results were seen for the ADG from birth to 21-day weight (99.6±7.0 vs. 153.6±6.0 g/d, P<0.0001; 119.6±6.7 vs. 133.6±6.9 g/d, P=0.8435), showing that LBW piglets grow at a consistent, albeit lower, rate than ABW conspecifics whether processed at d 1 or 3. Because of the lack of treatment by day interactions, it is most likely that the vocalization differences were a reflection of decreased LBW piglet vitality rather than decreased welfare in response to pain and distress.

Nest box sharing by Pekin ducks: a social network analysis

Makagon, Maja M. and Mench, Joy A., University of California, Davis, Animal Behavior Graduate Group and Department of Animal Science, One Shields Ave, Davis, CA, 95616, USA; jamench@ucdavis.edu

Social factors can play an important role in the selection of nesting sites by poultry. Using social network analysis, we examined the influence of the presence of pen-mates in the nest boxes on nest box use by Pekin ducks. Individually marked ducks (N = 32) were housed in groups of eight. Each group was given access to eight nest boxes (NB) at 18 weeks of age. Behavior was video recorded for 24 hours during weeks 26 and 30. During the period of highest NB use (03:00-07:00 h), all NB were entered at least once in the majority (75%) of pens. However, individual ducks typically entered only 4 or 5 of the NB (range 1-8). There was no consistency in the positions of the nest boxes entered across weeks, indicating that ducks did not form location preferences. On average, eggs were laid in only 3 (range 2-4) of the NB. Ducks typically nested alone or in groups of 2-3, but up to four ducks were observed to occupy a NB simultaneously. Within pens, there was substantial between-duck variation in both the number of NB entries per day (mean 12.0 ±2.2, range 1-38) and the proportion of entries into occupied NB (0.5 ±0.2, range 0.1-0.9). Standard deviations of the mean association indices within dyads, which were calculated using SOCPROG, were larger ($P<0.05$) than expected in all pens, indicating that ducks preferentially entered NB occupied by particular pen-mates. Nesting affiliations tended to be reciprocal, since there was a moderate but significant (r_s between 0.32 - 0.56; $P \leq 0.05$) correlation between the number of times each duck in a dyad entered a NB when the other member of that dyad was in that NB. These results demonstrate that the presence of particular pen-mates in NB attracts Pekin ducks to use those NB, indicating that social affiliation plays an important role in nest site selection.

Five-day headache after disbudding?

Hokkanen, Ann-Helena[1,2], Pastell, Matti[1,3] and Hänninen, Laura[1,2], [1]University of Helsinki, Research Centre for Animal Welfare, P.O. Box 57, 00014 University of Helsinki, Finland, [2]University of Helsinki, Production Animal Medicine, P.O. Box 57, 00014 University of Helsinki, Finland, [3]University of Helsinki, Agricultural Sciences, P.O. Box 27, 00014 University of Helsinki, Finland; ann-helena.hokkanen@helsinki.fi

Calves may suffer from pain for several days after disbudding. We aimed at studying the effect of 5-day pain medication on calves' behaviour. We performed a randomized double-blinded study on 10 one-month-old calves. The calves were disbudded (with the heat cauterisation) under sedation and local analgesia. Immediately before sedation, the calves were administered either oral ketoprofen (4 mg/kg) or placebo. The oral treatment (2 mg/kg) was continued over four more days and the calves' behavior was filmed simultaneously. We calculated the total duration of the observations (%), bout durations and frequencies for calves' resting behavior during one day before disbudding on the 2nd and 5th day after. The effect of treatment on calves' behavior was analyzed with a linear mixed effects model taking repeated measures into account. We found a significant interaction ($P<0.05$) between the treatment and days on number of lying bouts and a tendency on the length of the lying bout ($P=0.09$): Pain medicated calves had longer (51.1 ± 4.6 min vs. 37.8 ± 4.6 min) and fewer (20.8 ± 2.2 vs. 27.6 ± 2.2) lying bouts on the 5th day after disbudding ($P<0.05$ for all) than placebo treated calves. The disbudding of placebo calves had no effect on the percentage of the time the calves spent resting (76-78%, SE 2.4%) but it did increase medicated calves' resting from the 2nd to the 5th day after disbudding (72% vs. 80%, se 2.4%, $P<0.05$). We concluded that as five-day pain medication induced calves to lie down more than placebo treated calves, the calves most likely suffer from pain at least five days after disbudding.

Altered thermal but not mechanical nociceptive threshold in moderately lame broiler chickens

Caplen, Gina, Hothersall, Becky and Murrell, Jo, University of Bristol, Clinical Veterinary Science, Langford, North Somerset, BS40 5DU, United Kingdom; gina.caplen@bristol.ac.uk

Lame broiler chickens show altered behaviour patterns that suggest a reluctance to move, but the extent to which lameness is painful remains unknown. To attain a measure of pain sensitivity, we evaluated novel threshold testing devices (TopCat Metrology Limited) that apply a ramped thermal or mechanical stimulus to the leg (potentially painful 'test' site) or keel (reference site) of broilers. Baseline thresholds of moderately lame (Gait Score 3) and sound (Gait Score ≤0.5) broilers were compared by recording the temperature or pressure at which broilers made a clear behavioural response – usually weight-shifting – to the stimulus. The stimulus was then immediately removed. Following validation for repeatability and reproducibility, testing used thermal and mechanical probes applied to the right leg, and a thermal probe to the keel. For each site, 6 tests were conducted at 15 minute intervals and a mean value calculated for each bird. Results were compared with latency to lie (LTL) when placed in 4cm deep tepid water. Our experimental group yielded 6 GS≤0.5 birds and 13 GS3 birds. The mean thermal threshold (increase above initial skin temperature) was significantly lower in GS ≤0.5 birds (3.2 °C, SE = 0.4) than in GS3 birds (4.5 °C, SE=0.3) for the leg (t-test, t=2.70, P=0.016) and tended to be lower for the keel (GS ≤0.5, mean=6.7 °C, SE=0.5; GS3, mean=7.7 °C, SE=0.2; t=-2.00, P=0.065). Mean mechanical leg threshold (GS≤0.5=3.7N, SE=0.5; GS3=4.1N, SE=0.3) did not differ significantly between groups but was almost significantly correlated with thermal leg threshold (Pearson's r=-0.46, P=0.052). Mean LTL was significantly longer in GS≤0.5 (782s, SE=118) than GS3 (202s, SE=65) birds (Mann-Whitney U test, z=-3.10, P=0.001) and tended to be correlated with thermal leg threshold (Spearman's r=0.43, P=0.073). Results suggest GS3 birds avoided bearing weight, but further studies are required to separate the confounding effects of body mass and lameness.

Behavioural responses to disease in dairy calves

Jensen, Margit Bak[1], Trénel, Philipp[2], Svensson, Catarina[3], Decker, Erik L.[1] and Skjøth, Flemming[2],
[1]University of Aarhus, Faculty of Animal Science, Department of Animal Health and Bioscience,
Blichers Allé 20, DK-8830 Tjele, Denmark, [2]AgroTech A/S, Institute for Agro Technology and Food
Innovation, Udkærsvej 15, DK - 8200 Aarhus, Denmark, [3]Swedish University for Agricultural
Sciences, Faculty of Veterinary Medicine and Animal Science, Department of Animal Environment
and Health, P.O. Box 234, SE-532 23 Skara, Sweden; margitbak.jensen@agrsci.dk

Feeding behaviour and activity in responses to disease in dairy calves (N=158) were measured automatically via a milk feeder and leg-attached accelerometers. Calves (Danish Holstein, Danish Red and Jersey) were clinically examined three times weekly from 10 to 52 days old. Calves were housed in groups of 10 and assigned to either high milk allowance (8l/day for large breeds and 6.4l/day for Jersey) or low milk allowance (6.4l/day for large breeds and 4.8l/day for Jersey). Using linear mixed models, the behavioural variables were modelled as a function of (a) a categorised age variable, two categorised time variables describing (b) days prior and within a classified lung disease period and (c) days prior and within a classified diarrhoea period, (d) breed, (e) sex, (f) milk allowance, (g) age difference to oldest calf in group, as well as the first order interactions. Pen, calf within pen, and age category within calf and pen were modelled as random effects. The number of unrewarded visits was found to increase with age, most pronounced so in Jerseys (AxD, $P<0.001$), and was higher in calves offered a low milk allowance (F, $P<0.01$). Only Jerseys showed a significant decline ($P<0.001$) in unrewarded visits 1 to 2 days before a detected diarrhoea period when contrasted to unaffected levels. All calves showed a significant decline ($P<0.01$) in activity intensity while being upright 1 to 2 days before a detected diarrhoea period. In summary, three feeding behaviour variables (number of unrewarded visits, unrewarded visit duration, and actual-to-scheduled milk intake ratio) and two activity behaviour variables (activity while upright and number of postural changes) were found to change in response to a forthcoming disease. However, this often involved interactions with breed and/or milk allowance, indicating the complex nature of behavioural changes in response to disease.

Effects of sampling liver biopsies on dairy cow behaviour

Mølgaard, Lene[1], Damgaard, Birthe[2], Bjerre-Harpøth, Vibeke[2] and Herskin, Mette S.[2], [1]Aarhus University, Faculty of Science, Department of Biological Sciences, Ny Munkegade 114, DK-8000 Aarhus C, Denmark, [2]Aarhus University, Faculty of Agricultural Sciences, Department of Animal Health and Bioscience, P.O. Box 50, DK-8830 Tjele, Denmark; lenemoelgaard@hotmail.com

In dairy cows, sampling of liver biopsies is standard procedure for examination of liver metabolism. However, the impact of the procedure is poorly investigated. The aim of this study was to examine behaviour during and after sampling of liver biopsies. Data were collected from 16 cows using direct observations (5 hours) and IceTag activity monitors (22 hours) on Tuesdays and Wednesdays Weeks 4 and 3 before calving. The IceTags were attached one week before the experiment and registered number of leg movements and upright vs. lying position. Tuesdays, the cows were undisturbed. Wednesdays, the cows were head locked and blood sampled as part of another experiment focusing on metabolism, and in Week 3 subjected to percutaneus needle biopsy sampling after local anaesthesia (2% procaine). No analgesics were used. Data collected Tuesdays were used to rule out gestation week effects. Data collected the first Wednesday were used as control. Week of gestation did not show any effect. During the sampling restlessness (body movements) increased (paired t-test, 5±2 vs. 17±3; $P=0.008$), frequency of head movements increased (Signed Rank-test, 0 (0(25%)-0(75%)) vs. 0.5 (0-5), $P=0.016$), and duration of rumination showed a tendency for decrease (paired t-test; 111±31 vs. 27±21, $P=0.064$). After the sampling occurrence of behaviour typically connected to discomfort or pain (teeth grinding, leaning and tail pressing) increased (Signed Rank-test, 0 (0-0) vs. 2 (0-62), $P=0.039$) and comfort behaviour (licking and scratching) decreased (paired t-test, 21±11 vs. 17±5, $P=0.044$). There was no effect on activity or food and water intake. The results suggest that sampling of liver biopsies induces behavioural changes in dairy cows both during and after the procedure - especially behaviours previously associated with pain or discomfort. If this is confirmed in future experiments focusing on pain, we recommend improvement of the anaesthetic protocols and possible inclusion of analgesics.

Sickness behaviour in dairy cows challenged with *E. coli* induced mastitis

Jensen, Katrine Fogsgaard[1], Røntved, Christine M.[2], Sørensen, Peter[3] and Herskin, Mette S.[2],
[1]*University of Aarhus, Faculty of Science, Department of Biological Sciences, Ny Munkegade 114, DK-8000 Aarhus, Denmark,* [2]*University of Aarhus, Faculty of Agricultural Science, Department of Animal Health and Bioscience, P.O. Box 50, DK-8830 Tjele, Denmark,* [3]*University of Aarhus, Faculty of Agricultural Science, Department of Genetics and Biotechnology, P.O. Box 50, DK-8830 Tjele, Denmark; katrine@fogsgaards.dk*

Mastitis is a high incidence disease among dairy cows and considered to be painful. However, the consequences of mastitis in terms of animal welfare are not known. The aim of the present experiment was to examine behavioural changes in dairy cows before and after experimental challenge with intramammary *E. coli*. Effects of experimentally induced mastitis were examined in 20 Danish Holstein-Friesian cows kept in tie-stalls, milked at 6am and 5pm and fed a TMR based on maize silage for *ad libitum* intake. The cows were in 1st lactation and 3-6 wk after calving. After evening milking at t=0, each cow received an intramammary infusion with 20-40 CFU *E. coli* in one front quarter with <27,000 SCC/ml milk. Clinical examinations were conducted to confirm infection and SCC(x1000/ml milk) was 1,153.34±177.73, 3,936.64±264.86 and 3,237.8±348.51 for day 0, 1 and 2, respectively. Behaviour was recorded using video from day -2 to +2 after challenge, and only registered when the cows were not disturbed by humans. The behaviour on the different days was compared using mixed model analyses. The results show that time spent feeding was significantly lower ($F_{4,76}=4.69;P<0.01$) in the initial 24 hours(day 0) after infection compared to the other days(2.6h±0.3 vs. 3.5h±0.3, 3.3h±0.2, 3.6h±0.3, 3.3h±0.2 for days -2 to +2, respectively). The frequency of self-grooming behaviour dropped significantly on day 0(67.3±15.2;$F_{4,19}=4.00;P<0.05$) compared to day -2 (89.5±17.9;$P<0.01$), -1 (92.1±21.0;$P<0.01$) and +2 (96.6±21.1;$P<0.01$), and tended to be lower on day +1 (81.6±18.9;$P<0.06$). The duration of standing idle increased significantly on day 0 ($F_{4,70.4}=10.11;P<0.0001$). The results suggest that dairy cows show classic signs of sickness behaviour in the hours after challenge with *E. coli*. Changes in feeding time and time spend idle are suggested to be potential behaviours for future automatic health detection. Further research is needed in order to confirm the existence of pain.

Relationship between forced physical activity, leg health and performance in heavy male turkeys

Berk, Jutta, Institute of Animal Welfare and Animal Husbandry, Friedrich-Loeffler-Institut, Dörnbergstr. 25/27, 29223 Celle, Germany; jutta.berk@fli.bund.de

Leg disorders account for considerable problems in commercial turkey husbandry. This study investigated the effect of forced physical activity (300 m running once per day, 5 days per week) on leg health and performance of male turkeys (B.U.T. Big 6). Turkeys were placed in 6 floor pens (each 36 m^2, 124 animals). Group 1 (Control, n=248) received no training, group 2 (G2-8, n=248) was trained from week 2 until week 8 and group 3 (G2-21, n=248) was trained from 2 to 21 weeks of age. Turkeys were trained in groups with diminishing group sizes from 124 animals (wk 2-8) to 10 birds (wk 19). Traits recorded were body weight and food consumption (all birds) as well as weight and length of femora and tibiae (at a time 40 birds per group, wk 8 and 21). In the middle and at the proximal end of tibiae and femora total area, total density, cortical area, cortical density and strain strength index were measured by peripheral quantitative computer tomography. Additionally, bone breaking strength, walking ability (score 1=normal to 4=incapable of walking) and leg posture (normal, x-, o-, wide legs) were assessed. Data were analysed using the mixed procedure of SAS. No differences were found between groups for growth, mortality and feed intake. At week 21, leg posture was better in trained groups (1.74±0.04 vs. 1.60±0.04, $P<0.05$) compared to control birds (1.88±0.05, $P<0.05$). Furthermore, training improved walking ability in G2-21 (1.51±0.04, $P<0.05$) compared to control group (2.2±0.05) and G2-8 (2.2±0.05, $P<0.05$) and led to increased total density and higher breaking strength of femora ($P<0.05$). Results showed that positive effects on walking ability, leg posture and bone strength can be achieved with intensive physical training without negatively affecting performance traits. Preventing and managing skeletal problems in turkeys can improve health and welfare of birds.

Behavioural responses of juvenile Steller sea lions (*Eumetopias jubatus*) to hot-iron branding

Walker, Kristen[1], Mellish, Jo-Ann[2,3] and Weary, Dan[1], [1]University of British Columbia, Animal Welfare Program, 2357 Main Mall, Suite 180, Vancouver, BC V6T 1Z4, Canada, [2]University of Alaska Fairbanks, School of Fisheries and Ocean Sciences, Fairbanks, AK, 99775, USA, [3]Alaska SeaLife Center, Seward, AK, 99664, USA; walkerkr@interchange.ubc.ca

Marking and tracking methods used on wildlife may alter behaviour and cause pain. Hot-iron branding is a common procedure for marine mammals, but little is known about how animals respond to branding. We monitored the behavioural responses of juvenile Steller sea lions (n=11) for 3 d before (Pre-brand) and for each of the 3 d after (Days 0, 1 and 2) hot-iron branding. Six behaviours were predicted to respond to post-branding pain, including changes in locomotion, time spent in the pool, lying position, wound-directed grooming (scratching, biting and rubbing the affected area), time alert and time with pressure on branded side. Mixed model analysis was used to test the effects of hot-iron branding on the six behaviours. Four behaviours changed after branding. Specifically, the proportion of time sea lions spent in locomotion decreased from 0.07 (+0.02) Pre-brand to 0.03 (+0.01) on Day 0 (backtransformed means + SE; $P=0.044$), and returned to Pre-brand by Day 1. Wound-directed behaviour increased from 0.0 (+0.0) Pre-brand to 0.02 (+0.01) on Day 0 ($P=0.004$) and 0.01 (+0.01) Day 1 ($P=0.023$), returning to Pre-brand levels by Day 2. Time spent in the pool declined from 0.17 (+0.06) Pre-brand to 0.05 (+0.04) on Day 0 ($P=0.024$), and returned to Pre-brand by Day 1. The time sea lions spent with pressure on their branded side showed little change from the 0.08 (+0.03) witnessed at Pre-brand to 0.10 (+0.03) on Day 0 (N.S.), but decreased to 0.02 (+0.02) on Day 1 ($P=0.047$) and 0.01 (+0.01) on Day 2 ($P=0.013$). Our results illustrate that these behavioural responses are useful in assessing sea lion pain, and may be useful in future studies looking at the efficacy of analgesics in reducing pain after hot-iron branding.

Use of a runway test in demonstrating turkey muscle pain

Ylä-Ajos, Maria[1], Hänninen, Laura[1], Tuominen, Satu[1], Pastell, Matti[2], Lehtonen, Anu[2] and Valros, Anna[1], [1]University of Helsinki, Department of Production Animal Medicine, P.O. Box 57, 00014 University of Helsinki, Finland, [2]University of Helsinki, Department of Agricultural Sciences, P.O. Box 27, 00014 University of Helsinki, Finland; maria.yla-ajos@helsinki.fi

Turkeys have been intensively reared for increased growth rate. Several researches have reported serum creatine kinase (S-CK) activity and muscle damage to increase with age in modern turkeys. To investigate whether these damages cause pain to the animal, 34 turkey toms were acquired from a commercial farm and separated into two groups with identical S-CK. The groups were housed together and after familiarization period, the birds were trained to walk on a runway (7.5 m × 1 m). The runway ended when the birds had left a start box and reached a food reward (maximum 3 min). A double-blinded setup was used to study the effect of three-day pain medication on turkey performance on the runway. The medication started at the age of 16.4±0.5 weeks with either meloxicam (initial dosage 1.0 mg/kg iv, maintenance 0.5 mg/kg/d iv) or a similar volume of saline. The birds were tested on the runway on the morning before the first medication, and the following two days. Ethical aspects of the experiment were judged and approved by the Finnish animal experiment committee. The walking speed was compared between the medicated and placebo-treated birds with a mixed model, taking repeated samplings into account. The medicated turkeys did walk faster the first 1.3 meters in the runway on the third test day than on the first test day (50±1 mm/min vs. 112±1 mm/min, $P=0.023$) while there was no change in the walking speed of the placebo treated birds. However, the difference between the treatments did not remain beyond the first part of the runway. The mean S-CK-level of the birds was 14.7 U/ml (min 4.7, max 32.0). We concluded that runway test is a useful tool to measure the willingness of a turkey to walk. However, more research is needed and the system needs to be validated.

Domestic pigs' responses to conspecific stress-induced vocalisation

Düpjan, Sandra, Tuchscherer, Armin, Schön, Peter-Christian, Manteuffel, Gerhard, Langbein, Jan and Puppe, Birger, Leibniz Institute for Farm Animal Biology, Behavioural Biology/Genetics and Biometry, Wilhelm-Stahl-Allee 2, 18196 Dummerstorf, Germany; duepjan@fbn-dummerstorf.de

Recent research has shown that domestic pigs show differential vocal responses to stressors of different type and/or quality, providing detailed information not only on the caller's identity but also on its affective state. There are controversial anecdotic reports on whether or not stress-related vocalisations ('distress-calls') elicit arousal or even stress in conspecifics hearing them. The use of distress-calls as a communicative signal, and hence a welfare-relevant propagation of distress between individuals deserves closer evaluation. In this study, we examined both behavioural and physiological (heart rate/-variability) responses of 24 juvenile female pigs to conspecific distress-calls in a playback-experiment, broadcasting both calls elicited by physical restraint (DC) or an artificial sound (AS) on consecutive days. Behaviour was scored in two-minute intervals before, during, and after playbacks, while heart rate and heart rate variability were analysed in ten-second intervals. For statistical analyses, we used mixed effect models with stimulus type, interval, day, and their pairwise interactions as fixed factors, subject as repeated factor, and additional pairwise t-tests. Subjects showed orientation responses to both sounds as indicated by significant decreases of heart rate for 20 seconds after onset of playbacks (compared to directly before onset; DC: t=4.79, $P<0.001$; AS: t=3.26, $P<0.01$) based on vagal activation. The unfamiliar AS elicited stronger physiological responses than DC (average decrease of 14 vs. 10 heart beats per minute), most likely due to its novelty. Subjects vocalised less ('grunt') during DC than during AS (t=4.39, $P<0.001$), and when DC stopped heart rate decreased by 11 beats per minute, while they showed no response when AS ended. These results suggest recognition of the conspecific, familiar sound. In conclusion, we found evidence for social recognition of conspecific distress-calls in pigs, but emotion propagation through acoustic communication of distress could not be verified at least for unrelated receivers and needs further research.

Undernutrition in early to mid pregnancy causes deficits in the expression of maternal behaviour in sheep that may affect lamb survival

Dwyer, Cathy M.[1], McIlvaney, Kirstin M.[1], Coombs, Tamsin M.[1], Rooke, John A.[1] and Ashworth, Cheryl J.[2], [1]Scottish Agricultural College, Animal Behaviour and Welfare, Kings Buildings, West Mains Road, Edinburgh EH9 3JG, United Kingdom, [2]The Roslin Institute, Division of Developmental Biology, Roslin, Midlothian EH25 9PS, United Kingdom; cathy.dwyer@sac.ac.uk

Undernutrition throughout pregnancy reduces the expression of maternal behaviour in sheep. We investigated whether undernutrition during the first two-thirds of pregnancy affects the expression of maternal behaviour in two breeds of sheep. Primiparous Scottish Blackface (BF: n=78) and Suffolk ewes (S: n=76) were fed either 100% (C: n=72) or 75% (R: n=82) of maintenance energy requirements between d1-91 of pregnancy over 2 years. Maternal behaviour data were collected during the first 2 h after birth and a subset of 45 ewes was tested for their ability to discriminate their own lambs in tests at 6 and 12 h after birth. BF ewes were quicker to start grooming their lambs after birth ($P=0.003$), spent more time grooming (51.8% vs. 33.2%, s.e.d.=3.65, $P<0.001$) and made more low-pitched bleats ($P=0.01$) to their lambs than S ewes. There were no overall effects of nutritional treatment. When tested at 6 h after birth, 62% of BF ewes distinguished between their own and an alien lamb, whereas only 5.6% of S ewes did (Fisher's Exact $P<0.001$). R ewes were more likely to prevent their own lamb from sucking than C ewes (35.3% R vs. 0% C, Fishers Exact $P=0.042$). R ewes were also slower to approach their own lambs in a maze test at 12 h ($P=0.017$), and tended to spend less time oriented towards their lamb ($P=0.064$) or interacting with their lambs during the test ($P=0.10$). Significantly fewer R than C lambs were reared to weaning in year 1 ($P=0.022$), and R ewes were more likely than C ewes to fail to rear any lambs to weaning (7.1% R ewes vs. 1.4% C ewes, Fisher's Exact $P=0.089$). The data suggest that, although the onset of maternal behaviour was not affected by undernutrition, there were some deficits in maternal care which may influence lamb survival.

Effect of time and parity on the behaviour of dairy cows during the puerperal period

Mainau, Eva, Cuevas, Anna, Ruiz-De-La-Torre, José Luis and Manteca, Xavier, School of Veterinary Science, Universitat Autònoma de Barcelona, Campus de la UAB, Ed. Veterinària, Departament Ciència Animal i dels Aliments, V0-135, 08193, Bellaterra, Barcelona, Spain; eva.mainau@uab.cat

The objective of this study was to investigate the effect of time and parity on behaviour as a possible indicator of discomfort associated with calving. This experiment is part of a larger project aiming to assess the effects of the non-steroidal anti-inflammatory drug meloxicam on pain associated with calving. Fifty-three dairy cows of first to sixth parity with calving that did not require mechanical assistance were included. Cow behaviour was observed using video recordings from 2 days before to 2 days after calving. Position of the cow in the pen, posture, and head and back behaviours were observed at 10 minutes intervals. Head turning, kicking, self-grooming and exploratory behaviours were observed continuously during 15 seconds every 10 minutes. Moreover, the position of the calf at birth (head or back position) and the calving assistance (score 0: non-assistance; score 1: some assistance with an easy manual pull) were evaluated. Statistical analyses were carried out with the SAS software usingmthe GENMOD procedure. As calving approached, cows showed a higher percentage of active behaviour, stayed less time in the eating and drinking area and showed a higher frequency of head turning and exploratory behaviour ($P=0.0030$, $P=0.0042$, $P=0.017$ and $P=0.0003$). Tail-up behaviour was observed more frequently from -12h to +6h around calving ($P<0.001$). Heifers showed more kicking behaviour before calving and a tendency to perform more self-grooming behaviour after calving than multiparous cows (0.38 ± 0.050 versus 0.24 ± 0.021 times/hour; $p=0.004$ and 0.21 ± 0.019 versus 0.16 ± 0.012 times/hour; $P=0.08$, respectively). During the pre-calving period, cows with score 1 at calving showed a higher frequency of turning head, kicking and self-grooming behaviour than cows with score 0 at calving ($P=0.03$ and $P=0.007$ and $P=0.01$ respectively). Although further research is needed, these preliminary results suggest that some behaviour parameters could be possible indicators of pain and discomfort associated with calving.

Calving difficulty in dairy cows results in less vigorous calves but does not affect the onset of maternal behaviour

Barrier, Alice, Ruelle, Elodie, Haskell, Marie and Dwyer, Cathy, Scottish Agricultural College, Animal Behaviour and Welfare, SSW Building, Bush Estate, PENICUIK, EH26 0PH, United Kingdom; alice.barrier@sac.ac.uk

Vigour at birth as well as appropriate maternal care is crucial to the neonate's development. However, in sheep and beef cattle, dystocia results in low vigour of the offspring and can result in altered maternal care, probably as a consequence of exhaustion and pain. The objective of the study was to determinate whether dairy calves born from an assisted calving (human intervention needed) would have lower vigour than calves delivered normally and whether their dams would show delayed and impaired maternal behaviour. Twenty-eight calvings (twelve normal and sixteen farm assisted including 4 malpresentations of the calf) from Holstein cows were observed continuously for 3 hours after expulsion of the calf from videos collected on the SAC farm (Dumfries, UK) between November 2008 and 2009. Calvings were balanced between the two groups for parity of the dam (primiparous vs multiparous), their genetic line, sex and birth weight of the calf and calving season. Assisted calves were slower to first attempt to stand, achieve standing, walk and approach the udder (Kaplan-Meier; $P<0.05$). Differences between the groups in the median latencies were 21, 27, 73 and 83 min respectively. Assisted calves also tended to be less likely to stand and walk within the first 3 hours after birth (Fisher exact Test; $P<0.1$), spent more time lying on their flank (+11min; $P=0.019$) and had more frequent bouts of this behaviour (Kruskall-Wallis; $P=0.033$). Assisted dams did not take longer to sniff or lick the calf and performed as much licking ($P>0.05$) as unassisted dams. However large individual variability was observed within dam's behaviours. Our results suggest that dairy calves born following difficult calvings have lower vigour in the first three hours after birth than unassisted calves. This might have longer-term effects on the health and survival of the calves.

Non-nutritive nursings: one mechanism to control milk transfer in domestic pigs

Illmann, Gudrun, Špinka, Marek, Haman, Jiří, Šimeček, Petr and Šilerová, Jitka, Institute of Animal Science, Department of Ethology, Praha 10-Uhříněves, 10400, Czech Republic; Illmannova@vuzv.cz

In mammals, milk transfer from the mother to the young is the most important parental investment. In domestic pigs up to 30% of nursings episodes can be completely without milk transfer (non-nutritive nursings, NNNs). Here, we posit the hypothesis that NNNs are a way for the mother sow to down regulate milk transfer through a reduced number of nutritive nursings (NNs) when piglets demand nursings too frequently. We further propose that NNNs may be a part of an honest signaling system that enables the sow to provide more frequent NNs to those litters that can prove their need through very frequent nursing initiations. We predicted that (P1) with an increasing number of all nursings (ALL=including NNNs and NNs) the proportion of NNNs should increase. (P2) However, because NNNs are not free but costs time and energy for the mother, we also predicted that with an increasing number of ALL the number of NNs should also increase. A meta-analysis was applied on eight studies (n=203 sows) that recorded the number of NNs (n=8993) and NNNs (n=2208) in domestic pigs. The studies included different breeds, different time of lactation and housing conditions. We confirmed both (P1) that the proportion of NNNs increased when the nursings were more frequent ($P<0.0001$) and (P2) that the number of NNs increased when nursings were more frequent ($P<0.0001$). In combination, these results show that the cost of each additional nutritive nursing that the piglets would instigate increases non-linearly at an ever steeper slope. In conclusion our results suggest that NNNs are adaptive for the sow because they down regulate the number of NNs and consequently decrease the milk output; hence they prevent excessive investment in the current litter.

Comparison of growth and performance of dairy calves suckling mother, suckling foster mother, fed *ad libitum* milk replacer and fed conventionally
Thorhallsdottir, Anna Gudrun and Thorsteinsson, Bjorn, Agricultural University of Iceland, dep. Environment, Hvanneyri, 311 Borgarnes, Iceland; annagudrun@lbhi.is

The common practice in the dairy industry is to separate the calves from the cow within 24 hrs. of birth and feed restricted quantities of milk until weaning, often approximately 10% of the calf's BW. Studies indicate that with this quantity, the calves are not able to express their full growth potential for the first few weeks, which may have negative effects in the long run. The growth and performance of 32 calves from Icelandic native cattle, that were assigned to four different treatments (n=8), was followed for the first 8 weeks of life. The treatments were: 1. With mother and suckling; 2. With foster mother and suckling; 3. Housed and fed *ad libitum* milk replacer and 4. Housed and fed conventionally (10% of body weight - 4 litre/day). The mothers with their calves were kept with the herd grazing and brought in for milking twice a day. The two foster mothers were kept with the eight calves grazing, while the housed calves were kept in a group in a pen with straw and fed from artificial teats. During the first eight weeks, the mother calves grew on average 1021 (±120) g/day; the foster mother calves 863 (±168) g/day; the ad lib calves 724 (±173) g/day and the conventionally fed calves 487 (±34) g/day. As the calves that were given *ad libitum* milk replacer did not grow as well as calves that were with their mothers ($P=0.0008$) and foster mothers ($P=0.07$) it is concluded that the amount of milk is not the only factor affecting the growth and performance of the calves. Being outdoors, although variable weather, might have had an effect, as well as the sole existence of the mother or a substitute mother per se.

Effects of maternal presence and environmental enrichment on food neophobia of piglets

Oostindjer, Marije, Mas-Muñoz, Julia, Van Den Brand, Henry, Kemp, Bas and Bolhuis, J. Elizabeth, Wageningen University, Adapation Physiology Group, Marijkeweg 40, 6709 PG Wageningen, Netherlands; marije.oostindjer@wur.nl

Low feed intakes of weanling piglets may be increased by stimulating feed intake before weaning. We investigated whether the presence of the mother, and environmental enrichment, reduce food neophobia of piglets before weaning. Pigs were housed in barren 9 m^2 pens, or in enriched 18 m^2 pens containing wood shavings, straw, peat and branches (n=16 litters). At 25 days of age piglets were exposed to novel food (cheese and chocolate-covered peanuts, five pieces per pig) for 7 min either in the presence of the sow (half a litter), who was familiar with the food, or in her absence, out of visual, auditory or olfactory range (other half of litter). Data were analyzed with generalized linear mixed models including sow presence, enrichment, their interaction and batch as fixed effects, and pen as random effect. In the presence of the sow piglets ate more chocolate peanuts (3.5 ± 0.8) than in her absence (2.4 ± 0.7, P<0.05). Sow presence also increased the percentage of piglets sampling both food items (89 ± 7% vs. 70 ± 10%, P<0.05). Sow presence affected, particularly in barren housing, latency to explore the food (barren-absent: 117.4 ± 45.2 s; barren-present: 13.4 ± 1.9 s; enriched-absent: 104.7 ± 29.3 s, enriched-present: 25.5 ± 9.7 s, interaction P<0.001). Chocolate peanut consumption was higher in enriched piglets than in barren piglets (as the sow ate many food items, tested for sow-absent only: barren-absent: 3.4 ± 1.5 g; enriched-absent: 20.5 ± 9.9 g, P<0.001). The mother can be an important factor in reducing food neophobia, possibly by reducing overall stress levels or by social facilitation. However, environmental enrichment may also affect behavioural development in such a way that piglets show a higher acceptance of novel food types before and likely also after weaning.

Effects of positive versus negative handling during pregnancy on maternal behaviour in sheep

Hild, Sophie[1], Coulon, Marjorie[1], Andersen, Inger Lise[2] and Zanella, Adroaldo Jose[1,2], [1]Norwegian School of Veterinary Science, Department of Production Animal Clinical Sciences, P.O. Box 8146 Dep., 0033 Oslo, Norway, [2]Agricultural University of Norway, Department of Animal and Aquacultural Sciences, P.O. Box 5003, 1432 Ås, Norway; sophie.hild@nvh.no

While positive human-animal relationships can reduce fear reactions in farmed animals, rough handling is likely to be a source of stress. Stress occurring during pregnancy is even more critical as it may affect the developing foetus. The offspring can also be directly affected in the neonatal period by the behaviour of the dam. In sheep, better survival of the young is linked to better mother-young bonding. Here we investigated whether late-pregnancy stress, caused by negative handling of ewes (i.e. unpredictable treatment by humans including swift movements and shouting; n=8), for 10 min twice a day during the 5 weeks prior to parturition, affected their maternal behaviour differently from ewes that experienced gentle handling (i.e. predictable treatment including calm behaviour and soft talking; n=8) during the same period. All ewes had twin lambs. We predicted that negatively treated (NEG) ewes would provide more care to their offspring to protect them from a threatening environment compared to positively treated (POS) ewes. With an ANOVA, we tested the effect of positive vs negative handling during pregnancy on maternal behaviours at birth, and on their 'Follow Score' which measured at 24 hours the conflict between following the offspring when carried away by a human and avoiding the proximity with the human handler. Treatments showed no effect on ewes vocalisations or aggression towards the lambs but NEG ewes spent more time grooming their offspring ($P<0.05$) than POS ewes. POS ewes, on the other hand, had a higher Follow Score, i.e. they followed their lamb more closely ($P<0.01$). In conclusion, NEG ewes showed more care towards their lambs but were less likely to keep close proximity with them when carried away by a human, suggesting a greater fear towards humans. This shows that handling of ewes during pregnancy has an impact on subsequent maternal behaviours.

Modifying attitudes and behaviour towards dairy cattle by multimedia-based cognitive-behavioural intervention

Windschnurer, Ines[1], Boivin, Xavier[2], Coleman, Grahame[3], Ruis, Marko[4], Mounaix, Béatrice[5] and Waiblinger, Susanne[1], [1]University of Veterinary Medicine, Veterinärpl. 1, 1210 Vienna, Austria, [2]INRA, Theix, 63122, France, [3]Monash University, Wellington Rd, 3800 Vic, Australia, [4]Wageningen UR Livestock Res., P.O. Box 65, 8200, Netherlands, [5]Institut de l`Elevage, Monvoisin, 35652, France; Ines.Windschnurer@vetmeduni.ac.at

A multimedia training for stockpeople to improve human-animal interactions on cattle farms was tested for its effectiveness. Using cognitive-behavioural intervention, it was designed to improve attitudes towards cattle, handling behaviours, and, consequently, to reduce animal fear and stress. Austrian dairy cattle stockpeople were allocated to control (C) (9 farms, 9 people) or training (T) groups (10 farms, 14 people), balanced for the mean avoidance distance of their cow herds at visit1. During two farm visits, e.g., affective and behavioural attitudes towards cows, heifers, and calves were assessed with questionnaires (7-point Likert scale), human behaviour (tactile, vocal interactions) by behavioural observations during milking and milk feeding of calves, and cattle behaviour by avoidance and approach tests. T farms were visited before and after the training, C farms twice before the training. The reason of the study given to farmers was a repeatability analysis of a questionnaire and animal behaviour measures and to give feedback on a training program. Behavioural data were not available for all people (milking: C: n=8/T: n=11; milk feeding: C: n=3/T: n=4). Relative to C, T stockpeople displayed an improvement in behavioural attitudes towards cattle and handling them (Wilcoxon: median_score, (Min-Max): before T: 5.9, (4.7-6.9)/after T: 6.4, (5.7-7), Z=-2.73, $P<0.05$, n=14). An improvement in positive handling behaviours/cow (e.g., stroking, talking gently) during milking was found for T (Wilcoxon: before T: 0.7, (0.2-7.4)/after T: 1.0, (0.3-22.4), Z=-1.96, $P=0.05$, n=11). The behaviour towards calves did not change significantly ($P>0.05$). No training effect was observed on animal behaviour ($P>0.05$), maybe due to the short time between training and visit2. Nevertheless, the results demonstrate that this intervention program is a promising tool to improve the human-dairy cattle relationship. The study is part of the Welfare Quality research project, co-financed by the European Commission, contract no. FOOD-CT-2004-506508.

Environmental factors influencing responses of calves to human contact

Boivin, Xavier[1], Benhajali, Haifa[1], Castanier, Laure[1] and Egal, David[2], [1]INRA, UR1213 Herbivores, Site de theix, 63122 Saint-Genés Champanelle, France, [2]INRA, UE des Monts D Auvergne, Le Roc, 63210 Orcival, France; xavier@clermont.inra.fr

When cattle are reared outdoors, it is important to identify factors that influence responses of calves to humans. A possible effect of taming cows before calving was tested on the responses of their calves to human through maternal influences. About 2 months before calving, 30 outdoor reared Salers cows were tested twice for their response to a person using 3 different tests (docility test in a pen, avoidance distance at the feeding barrier, avoidance distance in the yard). Cows were allocated to two groups balanced according to previous test responses and calving day. 6 weeks before calving, one group was tamed (significant reduction of flight distance) for two weeks (90 min twice a day, standardised individual approach with food reward). From the age of 2 days on, the calves from both groups of cows were gentled for 5 min in isolation, once a day over 3 weeks. At the end of the gentling treatment and at 12 weeks of age, calves were tested in a human approach test. As weather conditions improved strongly during the calving period, daily temperatures (10 °C/25 °C), and humidity (17mm/0mm) were collected and relationship with calves' response to a person investigated. The responses of the cows were repeatable within test (r_s>0.57, P<0.05)but not between tests. Variance/covariance analyses revealed that neither responses of cows (by contrast to a similar previous study) nor taming them explained the responses of calves to humans (P>0.05). But at three weeks of age, stroking the animals during the test was related to temperature (α=2.94, F=5.61, P<0.05) and humidity (α=6.21, F=16.22, P<0.001), when running duration was also related to humidity (α=2.21, F=9.23, P<0.001). In this study, taming cows before calving seemed unsuccessful for influencing their calves' behaviour. But our results suggest that weather conditions could have affected test results and masked other influential factors. The authors wish to thank the French region Auvergne within the project 'Territoire, Agriculture, Alimentation en Auvergne' for funding this study.

Investigating greeting behaviour in dogs reunited with a familiar person

Rehn, Therese and Keeling, Linda J., Swedish University of Agricultural Sciences, Department of Animal Environment and Health, P.O. Box 7068, 750 07 Uppsala, Sweden; Therese.Rehn@hmh.slu.se

Reunion between a dog and a known person is thought to be experienced as a positive event by the dog, and may thus lead to expression of behaviours indicative of a positive emotional state. The aim with this study was to identify these behaviours and to investigate the effect of different types of interactions on dog behaviour around reunion with a familiar person. Behavioural and physiological responses of twelve female beagles were measured at reunion, but only the behavioural data is reported here. Three different treatments were applied and each dog experienced all three treatments in a balanced design; the familiar person initiated physical and verbal contact in a calm and friendly way (PV), there was verbal contact only (V), or the person ignored the dog (I). We aimed to vary the interaction intensity initiated by the human in our chosen treatments. Interaction continued for 90 seconds during which the person behaved in a standardized way according to treatment. Lip licking was recorded continuously and other behaviours every 5 seconds. Dogs showed more lip licking during interaction in PV than in the other treatments (PV: 1.43±0.1; V: 0.44±0.05; I: 0.39±0.05 (mean±SE), χ^2=11.45, $P<0.01$), indicating that close physical contact elicited more reinstatement behaviour. They performed least body stretching in V (PV: 0.05±0.02; V: 0.02±0.01; I: 0.05±0.01, χ^2=6.0, $P<0.05$), vocalized most (PV: 0.11±0.02; V: 0.23±0.03; I: 0.08±0.02, χ^2=6.28, $P<0.05$) and tended to yawn more. Being the usual type of reunion, we assume that PV was experienced as most positive by the dogs and therefore suggest that lip licking, which has previously been related to negative arousal, may also be an indicator of positive arousal. We also propose that the V treatment, since it was an unusual form of reunion, may have elicited mild conflict/attention seeking behaviours.

When inactivity predicts poor reproductive success in farmed mink, is poor welfare the link?

Meagher, Rebecca, Hunter, Bruce, Bechard, Allison and Mason, Georgia, University of Guelph, Animal and Poultry Science, 50 Stone Road E., Bldg. 70, Guelph, ON N1G 2W1, Canada; rmeagher@uoguelph.ca

Individual mink range from being very stereotypic to extremely inactive, these responses being inversely correlated. In Denmark, stereotypic/less inactive phenotypes have been statistically associated with increased litter-sizes, and reduced fear and baseline cortisol. We aimed to replicate these findings on 3 farms in Ontario, and test the hypothesis that inactive females' impaired reproduction reflects higher stress (the alternative being that it is caused by excess body-fat). For 556 Black and Pastel mink, we recorded pre-breeding activity levels; litter-size and other reproductive variables (prevalence of barrenness; kit survivorship and growth); faecal corticosteroid output; fear in 'glove tests'; and body-fat scores. As expected, inactivity showed a trend to predict small litters in reproductive females ($F1,459=3.02$; $P=0.08$), and was associated with greater body-fat ($F2,491=25.99$; $P<0.0001$). However, inactivity did not predict fearfulness, and unlike previous studies tended to predict lower corticosteroid output ($F1,61=3.59, P=0.06$). Furthermore, inactivity predicted improved production in other ways: it correlated positively with kit growth (controlling for litter-size, female kits: $F1,404=22.83$; males: $F1,398.2=36.57$; $P<0.0001$), and compared with very active females, very inactive mink were less likely to be barren (Fisher's Exact test; $P=0.06$). The one link between reproduction and stress occurred in Pastels: fearfulness predicted greater risk of barrenness (Fisher's $P=0.006$). However, this was independent of activity levels. Thus stress seemed unlikely to cause the small litters of inactive females. Instead, excess body-fat seemed responsible: in Pastels, body-fat tended to inversely predict litter-size (Fat × colour $F2,475=3.45$, $P=0.03$; Pastels: $F2,219=2.50$, $P=0.08$); and across all the mink, statistically controlling for body-fat abolished the apparent relationship between inactivity and litter-size. Overall, fatter bodies, not stress, thus reduce the litter-sizes of inactive mink, at least on Ontario farms. In failing to replicate Danish findings, our data also suggest that the correlates of having a stereotypic versus an inactive phenotype may vary between mink populations.

Belief in animal sentience during veterinary education

Clarke, Nancy, Main, David and Paul, Elizabeth, University of Bristol, Animal Behaviour and Welfare Group, Clinical Veterinary Science, Langford, BS40 5DU, United Kingdom; n.clarke@bristol.ac.uk

Research suggests that en route to graduation, veterinary students may decreasingly view animals as having capacity for conscious mind or sentience. Such attitudes may underpin a decrease in the likelihood to treat animals for pain, or to apply other welfare-relevant interventions. Consequent disparities in the recognition and subsequent management of animal pain and well-being in the clinical setting may have significant animal welfare implications. We developed a BIAS (Belief In Animal Sentience) questionnaire to investigate sources of variation in beliefs about animal sentience during veterinary education. To explore the effects of gender and pet ownership experience on sentience beliefs, two cohorts of first-year veterinary students (n=193) at one British university completed the BIAS questionnaire in 2008 and 2009. To study the possible effects of veterinary education, data was collected from a cohort of veterinary students during their final years of study in 2008 (n=83), having first completed the BIAS questionnaire during their first year of study in 2004. Veterinary student's belief in animal sentience varied according to animal species, increasing across the phylogenetic scale. Prior pet ownership was related to sentience beliefs. For example, students with experience of owning a cat gave significantly higher sentience scores to cats (F=4.291, $P<0.05$). Compared to male students, female students attributed significantly higher sentience scores for dogs (F=12.470, $P<0.005$), sheep (F=16.489, $P<0.001$), rabbits (F=5.996, $P<0.05$), lions (F=4.529, $P<0.05$), chickens (F=3.850, $P=0.05$) and cats (F=7.252, $P<0.01$). Contrary to previous findings, no significant changes were found in belief in animal sentience across veterinary education, although a trend indicating an increase in pig sentience reached near significance (T=1.824, $P<0.072$). Our findings offer insight into how gender, pet ownership, cohort and veterinary education may relate to belief in animal sentience and subsequent attitudes towards pain assessment and management in different animal species.

A comparison of the effects of three dehorning methods on the welfare of *Bos indicus* calves

Sinclair, Stephanie[1], Petherick, J Carol[2] and Reid, David[2], [1]Beef Genetic Technologies CRC, CSIRO Livestock Industries/The University of Queensland, P.O. Box 5545, CQ Mail Centre Qld 4702, Australia, [2]Beef Genetic Technologies CRC, Agri-Science Queensland, P.O. Box 6014, Parkhurst Qld 4702, Australia; stephanie.sinclair@csiro.au

In the northern Australian beef cattle industry a variety of instruments are available for dehorning and dehorning without anaesthesia/analgesia is permissible. Forty-four Bos indicus calves (2 - 6 months old, mean liveweight 135 kg) were allocated to four groups (n=11/treatment; animal ethics approval RH255/08). The responses of calves to dehorning by three instruments (scoop dehorners (Sc); dehorning knife (K); hot-iron dehorner (HI)) were compared to those of polled calves sham-dehorned (C). Vocalisation counts and number of struggles during dehorning were compared using a Generalised Linear Model. Plasma cortisol response and liveweight changes post-treatment for the four treatments were compared using ANOVA. During dehorning, mean number (± s.e.) of vocalisations indicated the HI group vocalised less ($P<0.05$; 4.0 ± 1.2) than the Sc (9.8 ± 1.9) and K (10.0 ± 1.9) groups. There were no treatment differences in the number of struggles. There were no treatment differences in the mean plasma cortisol response prior to dehorning. At 30 min post-treatment, cortisol levels of all three dehorning treatments (Sc=64.6nmol/l, K=62nmol/l, HI=62.2nmol/l) were greater ($P<0.05$; s.e.d. 6.24) than C (45.7nmol/l) and 5 hours post-treatment, Sc (57.3nmol/l) and K (61.3nmol/l) were greater ($P<0.05$; s.e. 7.72) than HI (39.7nmol/l) and C (38.8nmol/l). Liveweight changes were similar ($P>0.05$) among all treatment groups at 4 weeks post-treatment (159kg ± s.e. 1.9). The effectiveness of the dehorning instruments was assessed 3 months post-treatment. HI was the least effective with 15 (of 22) horns having regrown. In comparison, 4 (of 22) had developed regrowth or partial scurs in Sc group and 5 (of 22) in K. It was concluded that hot-iron dehorning may cause less stress and pain compared to using scoop dehorner or knife, but as it was largely ineffective as more animals would require dehorning again, which would be detrimental to their welfare.

Welfare impact of foot disorders in dairy cattle

Bruijnis, Mariëlle R. N.[1], Beerda, Bonne[2] and Stassen, Elsbeth N.[1], [1]Wageningen University, Department of Animal Sciences, Adaptation Physiology Group, Animals and Society, P.O. Box 338, 6700 AH Wageningen, Netherlands, [2]Wageningen University, Department of Animal Sciences, Adaptation Physiology Group, P.O. Box 338, 6700 AH Wageningen, Netherlands; marielle.bruijnis@wur.nl

Foot disorders are an important health problem in dairy cattle, in terms of animal welfare and economics. Objective of this study is to quantify welfare impact of different foot disorders, both clinical and subclinical. The welfare model uses the same structure as an earlier model describing economic consequences of foot disorders, with input being derived from literature and experts. Welfare impact is assessed taking into account incidence, duration and severity of foot disorders. Foot disorder severity is based on locomotion score (five-point scale). Locomotion score is a widely accepted parameter that assumingly mirrors estimated pain intensity (1=slight...5=severe), which was based on pathology of the foot disorder and its impact during activity. Foot disorders have an impact on the fulfilment of behavioural needs of dairy cows, like resting and gaining access to resources by moving around. Behavioural impact was included in the model by determining the weighing of severity against duration. This weighing was benchmarked against expert opinions. The model quantifies welfare impact of different foot disorders in terms of severity and duration. Behavioural impact of foot disorders seemed significant only when locomotion scores were higher than two, which supported a non-linear relationship between foot disorder severity and welfare. Digital dermatitis (pain intensity 3.9) has relatively high clinical incidence and duration and therefore highest welfare impact. Foot disorders that mainly occur subclinical, sole haemorrhage (pain intensity 1.3) and interdigital dermatitis/heel erosion (pain intensity 1.5), have a substantial impact on welfare because of long duration. Interdigital phlegmon (pain intensity 4.7) has lowest welfare impact due to short duration. The model facilitates a transparent evaluation of welfare impact of different foot disorders. Gaps in knowledge become clear, spurring further research. The economic model and welfare model aid in approaches to improve dairy cow foot health.

Positive affect and learning: exploring the 'Eureka Effect' in dogs

McGowan, Ragen T.S., Rehn, Therese, Norling, Yezica and Keeling, Linda J., Swedish University of Agricultural Sciences, Department of Animal Environment and Health, SLU, Uppsala, Sweden; Ragen.Trudelle-SchwarzMcGowan@rdmo.nestle.com

Animals may experience positive affective states in response to their own achievements during problem solving (e.g. 'Eureka Effect'). We aimed to investigate such responses in dogs and to separate these from reactions to rewards per se using a yoked control design. We also investigated whether the intensity of response would vary with different rewards. Twelve female beagles were assigned to matched pairs and served as both experimental and control animals at different stages of the experiment. We trained all dogs to perform distinct operant tasks and exposed all dogs to additional operant devices without training (controls). One week later the dogs were tested in a new context where performance of the learned task was necessary to access one of three rewards: (1) food, (2) social-human and (3) social-dogs. For experimental dogs, reward access was granted immediately upon completion of the task. For controls, access was granted after a delay equal to their matched partner's latency to complete their task and was independent of any action taken by the dog. Thus, any detectable differences between the two groups could be attributed to the experimental animals having the opportunity to control their own access to the reward. During problem solving opportunities experimental dogs were more active (mean±SE; 2.63±0.08 steps/s vs. 2.29±0.16 steps/s; $F_{1,11}=37.36$, $P<0.0001$) and wagged their tails more (2.70±0.01 wags/s vs.1.21±0.08 wags/s; $F_{1,11}=205$, $P<0.0001$) than controls. The intensity of this response was influenced by the type of reward available after successful completion of the problem solving task with experimental dogs showing more activity ($t_{22}=3.41$, $P=0.0067$) and tail wagging ($t_{22}=3.02$, $P=0.0063$) when expecting a food reward over social contact with other dogs. Our results suggest that dogs react emotionally to problem solving opportunities and that providing opportunities for successful completion of problem solving tasks could stimulate positive affective states in dogs.

Coping with cognitive challenge is emotionally positively evaluated by pig

Zebunke, Manuela, Langbein, Jan, Manteuffel, Gerhard and Puppe, Birger, Leibniz Institute for Farm Animal Biology (FBN), Behavioural Physiology, Wilhelm-Stahl-Allee 2, 18196 Dummerstorf, Germany; zebunke@fbn-dummerstorf.de

Sensory and cognitive hypostimulation as well as the lack of environmental predictability and control can impair animal welfare (allostasis). The aim of the present study was to demonstrate that cognitive enrichment in form of food-rewarded learning tasks positively affects animals in the long term. Twenty four domestic pigs in groups of four were conditioned to individual acoustic signals on so-called call-feeding-stations. Over the course of 6 weeks, pigs were called repeatedly by its individual sound to visit the station within 3 min to get a food reward. Additionally, the animals were set on an increasing fixed ratio button-push schedule. Each pig was called between 28 and 33 times per day and could meet its feeding requirement with 80% correct performance. Simultaneously, 24 conventionally fed control pigs were housed in an adjacent pen. Differences in the behavioural and physiological reactivity between experimental and control animals were analysed in a repeated open field/novel object test. Additionally, changes of resting heart rate/ -variability of the animals in the home pen were measured repeatedly. The animals subjected to the conditioning paradigm coped well with the cognitive challenge. During resting, a general autonomic activation, i.e. sympathetic (arousal) as well as vagal (positive emotional valence), evident only in the experimental animals, indicated positive emotional learning. In the behaviour tests experimental animals behaved more explorative and less fearful compared to control (contact with unknown object, 5.8 ± 0.3 s vs. 4.6 ± 0.3 s, $P<0.01$; defecation, 39.5 ± 3.2 s vs. 52.9 ± 3.2 s, $P<0.01$). Based on these results, we conclude, that pigs cope emotionally positively with adequate cognitive challenges. This kind of environmental enrichment provides pigs with environmental predictability and control, which leads to enhanced coping with challenging situations and has the potential to improve animal welfare in the long run.

Cage-mate/non-cage-mate categorisation of social odours in rats

Jones, Samantha[1], Burman, Oliver[2] and Mendl, Michael[1], [1]University of Bristol, Animal Welfare and Behaviour Group, Langford, BS40 5DU, United Kingdom, [2]University of Lincoln, Department of Biological Sciences, Riseholme Park, Lincoln, LN2 2LG, United Kingdom; s.m.jones@bristol.ac.uk

The development of cognitive categories of 'group-mate' and 'non-group-mate' likely underpins the occurrence of aggressive and affiliative behaviour towards 'in-group' and 'out-group' individuals. We investigated whether rats were able to categorise conspecifics as 'cage-mate' (CM) and 'non-cage-mate' (NCM). Forty-eight female Lister-hooded rats were housed in groups of twelve in four large enriched-cages. Four test rats, one from each cage, were trained to perform a two-choice olfactory discrimination task using odour cues collected on sand-filled bowls. Odour was collected from an individual CM and NCM and one odour was randomly allocated as the 'positive' (i.e. rewarded) stimulus. The subjects were then trained to discriminate between the two odours. Digging in the bowl containing the 'positive stimulus-odour' resulted in a food reward. No reward was given for digging in the other ('neutral') odour. A criterion level of performance (≥ 7 correct trials in three consecutive sessions of 10 trials each, $P=0.005$) was used to examine whether the rats were able to learn this discrimination. All rats reached criterion using four different sets of individual CM and NCM odours and the time taken to reach criterion decreased across replicates (Friedman(3)=7.816: odour-sets1-4, $P=0.05$). Rats then underwent trials in which the reward contingency was reversed. We predicted that if they had simply learnt to rapidly identify the rewarded exemplar, they should perform equally well in these reversal-trials as in preceding ones. However, if they had learnt the categorical discrimination, the reversed contingency (e.g. neutral-odour now rewarded and positive not) would result in initial errors, and more trials required to reach criterion. The latter prediction was supported (number of sessions to criterion increased to that of the initial learning set (Wilcoxon:odour-set4 vs. reversal:Z=-1.826, $P=0.068$; odour-set1 vs. reversal:Z=0, $P=1.00$). The results offer insight into how animals may categorise conspecifics, furthering our understanding of the social structure of group-housed animals.

Does human handling affect cognitive bias in dogs?

Reefmann, Nadine, Norling, Yezica and Keeling, Linda, Swedish University of Agricultural Sciences, Department of Animal Environment and Health, P.O. Box 7068, 75007 Uppsala, Sweden; Nadine.Reefmann@hmh.slu.se

The emotional state of an animal can be greatly affected by the way in which it is handled. Rough handling by a human is likely to increase negative emotions whereas gentle handling can lead to positive emotional states. The aim of this study was to compare emotion-related cognitive bias responses in dogs handled roughly and gently over a short period of time. Twelve beagles were trained to visually distinguish a rewarded box (food) and an unrewarded box (no food), as indicated by their running latency towards them. After successful training, half of the dogs were subjected to up to ten minutes of rough or gentle handling before the cognitive bias testing where the handler was also present. An unfamiliar person, acting as the rough handler, led the dog on a short leash and otherwise ignored it. The gentle handler was familiar and cuddled and played with the dog. Handling-induced cognitive bias was assessed by observing the dogs' behavioural reactions to a visually ambiguous probe box (colour intermediate to those of rewarded and unrewarded box). Data were analysed using linear mixed models. Overall, dogs ran faster to the rewarded (8 ± 2s (mean$\pm$SD)) than to the unrewarded boxes (26 ± 7s). All roughly-handled dogs reached the ambiguous probe box whereas only half of the gently-handled animals ran to the probe ($F_{1,10}=5.0$; $P<0.05$). Before running, roughly-handled beagles also tended to glance longer in the direction of the probe than gently-handled individuals ($F_{1,10}=4.1$; $P\leq0.07$). In conclusion, our data indicate that even short-term differences in the quality of human handling can influence cognitive bias responses in dogs. In contrast to cognitive bias studies on long-term emotional states, our results suggest that dogs in a negative emotional state seemed to appreciate a potential (food) reward more than those in a presumed positive emotional state.

Administration of serotonin inhibitor p-Chlorophenylalanine induces pessimistic-like judgement bias in sheep

Doyle, Rebecca E.[1,2], Hinch, Geoff N.[2], Fisher, Andrew D.[3], Boissy, Alain[4], Henshall, John M.[1] and Lee, Caroline[1], [1]CSIRO Livestock Industries, Locked Bag 1, Armidale NSW 2350, Australia, [2]School of Environment and Rural Science, University of New England, Armidale NSW 2351, Australia, [3]Faculty of Veterinary Science and Animal Welfare Science Centre, University of Melbourne, Melbourne VIC 3010, Australia, [4]INRA, UR1213 Herbivores, F-63122 St-Genès-Champanelle, Australia; rdoyle@csu.edu.au

Judgement bias has potential as a measure of affective state in animals. The serotonergic system may be one mechanism involved with the formation of negative judgement biases. It was hypothesised that depletion of brain serotonin would induce negative judgement biases in sheep. A dose response trial established that 40 mg/kg of p-Chlorophenylalanine (pCPA) administered to sheep via intraperitoneal injection for 3 days did not affect feeding motivation or movement required for testing judgement biases. Thirty Merino ewes (10 months old) were trained to a spatial location task used to test for judgement biases. Sheep learnt to approach the bucket when it was placed in one corner of the testing facility to receive a feed reward (go response), and not approach it when in the alternate corner (no-go response) to avoid a negative reinforcer (exposure to a dog). Following training, 15 sheep were treated with pCPA (40 mg/kg daily for 5 days) as described above. Control sheep were injected with sterile water in the same way. Treated and control sheep were tested for judgement bias 16 h after receiving the treatment for 3 and 5 days, and again 5 days after cessation of treatment. Testing involved the bucket being presented in ambiguous locations between the two learnt locations, and the response of the sheep (go/no-go) was measured. Following 5 days of treatment, pCPA-treated sheep approached the most positive ambiguous location significantly less than control sheep, suggesting a pessimistic-like bias (60% vs. 85% approaching more positive ambiguous location respectively; treatment x bucket location interaction $F_{1,124.6}=49.97$, $P=0.011$, tested by GLMM). No significant interaction was seen on day 3 of testing ($P=0.867$), nor 5 days post treatment ($P=0.068$). These results suggest that brain serotonin levels could be influential in the formation of judgement biases in sheep.

Investigating the reward cycle in chickens
Seehuus, Birgitte and Blokhuis, Harry, Swedish University of Agricultural Sciences, Department of Animal Environment and Health, P.O. Box 7068, SE-75007 Uppsala, Sweden; Birgitte.Seehuus@hmh.slu.se

The hypothesis of the reward cycle proposes that positive emotions are related to three motivational states: (1) an appetitive motivational state linked to the seeking/wanting of a resource, (2) a consummatory motivational state linked to sensory pleasure and (3) a post-consummatory motivational state associated with 'satiety/relaxation'. The goal of this study was to design an experimental set-up to allow manipulation of the different motivational states of 'the reward cycle' as a functional model for positive affective states in animals. A pen was designed to separate the different state related behaviours of a possible food related 'reward-cycle' in chickens. The pen had three areas; one with litter (1), one with food/water (2) and one covered, darker area, with heating lamp and perches (3). Two batches of 6 chickens in 3 pens, gave a total of 36 chickens. From three weeks of age the chickens were filmed for 11 hours a day. The videos were analysed (recording behaviour and area for each bird every minute) using the observational program, Interact. To validate the pen a Pearson's χ^2 test was performed. The chickens performed more exploratory behaviour (χ^2 (2, N=11,850 (number of total behaviours)) = 4,969.99 $P<0.0001$) in the appetitive part of the pen and preened ($\chi2=$ 1,904.18, $P<0.0001$) and rested ($\chi^2 = 1,097.34$, $P<0.0001$) more in the post-consummatory part. There were no significant differences between the pens. This supports the link between behaviours and specific areas and gives us the possibility to use the pen to manipulate the phases of the 'reward-cycle', and putatively related affective states. In further experiments we will study behavioural and cognitive effects of such manipulations as indicators of changes in positive affect. Data concerning the patterns of behaviours are under analysis.

Early-life-experiences alter attention to potential threats in sheep

Clark, Corinna C.A., Friel, Mary, Gamwells, Tom, Gordon, Amy, Murrell, Jo C. and Mendl, Michael T., University of Bristol, Department of Clinical Veterinary Science, Langford, BS40 5DU, United Kingdom; corinna.clark@bristol.ac.uk

Lambs commonly experience stress and pain from disease and practices such as castration and tail-docking; which, through early-life-programming, may result in the development of negative affective states (e.g. anxiety) and related behaviour (e.g. increased attention to threatening or ambiguous stimuli). We investigated whether early experience exerts long-term effects on how threatening sheep perceive the normal farming environment (e.g. being rounded up for husbandry procedures) to be. We used observations of natural threat perception behaviour, hypothesising that ewes with negative early-life-experiences would show heightened levels of threat monitoring, as measured by vigilance (scanning and fixed-gaze staring). As neonates 27 ewes were randomly assigned to treatments: CONT (handled); LPS (mild immune challenge aged 2 days); or TD (tail-docked using rubber ring aged 3 days). Approximately 16 months later they were divided into 3 groups of 9 ewes, comprising 3 ewes from each treatment. Groups were rotated between 3 field-plots of differing vegetative cover over 3 weeks. Behavioural observations consisted of: undisturbed behaviour; following rounding-up; following potentially threatening (dog) or non-threatening/ambiguous (bicycle) cues, with or without prior rounding-up. Data were analysed using multi-level statistical modelling (MLwiN). Treatment, cue, field, rounding-up-stressor, group and replicate were all found to be important variables. In the 25 minutes following cue presentation vigilance was more frequent following the dog (mean frequency = 7.7) than the bicycle (5.1); LPS ewes were more frequently vigilant (mean = 7.9, P= 0.027), TD ewes were least frequently vigilant (4.7) and controls intermediate (6.7). Our findings indicate that, in addition to the short-term welfare implications of early-life stressors, long-term alterations to emotional states should also be considered.

Cognitive abilities of farm animals: 'Stupid goats' (*Capra hircus*) are able to form open-end categories

Langbein, Jan and Meyer, Susann, Leibniz-Institute for Farm Animal Biology, Wilhelm-Stahl-Allee 2, 18196 Dummerstorf, Germany; langbein@fbn-dummerstorf.de

Categorization is regarded as the process of lumping together objects which share physical or conceptual relations. We studied the ability of dwarf goats to form open-end categories based on perceptual similarity of artificial stimuli. By applying a fully automated learning device, integrated in the animals' home pen, 26 goats were serially trained on eight visual 4-choice discrimination problems. Thereby, the positive (rewarded) category consisted of eight black shapes with an open center, while the negative category contained the same shapes but filled. Each problem ran for five days. To check for preferences for single shapes, we conducted a 1-day pre-test for each problem with all trials equally rewarded. Beginning from problem four, animals started to show ad hoc preferences for the positive shapes ($F_{7,168} = 105.52$, $P<0.001$) and the number of trials to learn the problems decreased significantly ($F_{7,168} = 99.39$, $P<0.001$), so it seemed that the goats had established the categories after training of only three problems. We conducted a transfer test to determine if the goats were able to generalize knowledge about the learned categories to new problems. Animals were given a mixed series containing all previously learned problems, in which 16 new problems were interspersed, that consisted of new shapes belonging to the same categories. Only the performance in the very first trial of each new problem was analyzed to show generalization abilities. A high overall learning success for all new problems (70.9%) on first-trial presentations indicate that dwarf goats selected shapes according to previously established categories. These results suggest that goats can form open-end categories based on shared perceptual similarity of stimuli and generalize this knowledge to new problems. The knowledge about advanced cognitive abilities of farm animals is the basis for future integration of challenging tasks into housing as new forms of cognitive enrichment.

Muddy conditions reduce lying time in dairy cattle

Tucker, Cassandra B.[1], Ledgerwood, David[1] and Stull, Carolyn[2], [1]University of California, Department of Animal Science, One Shields Ave, Davis, CA 95616, USA, [2]University of California, School of Veterinary Medicine, One Shields Ave, Davis, CA 95616, USA; cbtucker@ucdavis.edu

Dairy cattle spend less time lying down in rainy and windy conditions, but it is unclear if these behavioral changes are due to muddy conditions underfoot and/or exposure to inclement weather. To understand the effect of muddy conditions, we compared the behavioral response to dirt pens with 3 levels of soil moisture: 90, 74 or 67% dry matter (DM). Soil moisture was manipulated by adding water and fecal material to the top 15 to 20 cm of dirt in an 18x2.4m pen with a 3×2.4-m concrete apron in front of the feed bunk. Both the 74 and 67% DM treatments resulted in muddy conditions. Eight pairs of dairy cattle were exposed, in a balanced order, to each treatment for 5 d. On average, lying time was reduced by more than 3 h/d in very muddy conditions (12.6, 11.6, 9.0 h/d for 90, 74 and 67% DM; $P<0.001$). This difference was particularly marked on the first day of exposure when cows on the 67% treatment spent only 4 h/d lying, compared to 12.5 h/d when on dry soil (90%). Cows in the 67% treatment preferred to lie on concrete, while cows housed on dry soil never laid down on in this area (0, 1.3, 5.8 h/d lying on concrete in 90, 74, and 67% DM; $P<0.001$). Cows limited the surface area exposed to muddy conditions while lying (10, 10 and 18% of lying with front legs tucked and hind leg touching the body for 90, 74 and 67% DM; $P=0.058$). This change may be attributed to the thermoregulatory response to muddy conditions; however, body temperature did not differ between treatments. These results indicate that the muddy lying conditions (74 and 67% DM) for dairy cattle are less desirable that concrete flooring, reduce lying time and increase lying positions associated with potential heat loss.

The effect of TMR on dairy cow preference to be indoors or at pasture

Charlton, Gemma[1,2], Rutter, S. Mark[1], East, Martyn[2] and Sinclair, Liam[1], [1]Harper Adams University College, Animal Science Research Centre, Edgmond, Newport, Shropshire, TF10 8NB, United Kingdom, [2]Reaseheath College, Agriculture Department, Nantwich, Cheshire, CW5 6DF, United Kingdom; gcharlton@harper-adams.ac.uk

Grazing is considered a 'normal' behaviour for dairy cattle, although they may not be able to meet their nutritional requirements from grazing alone, and so require access to a total mixed ration (TMR). This study investigated whether offering a TMR at pasture affected cow preference to be indoors or outdoors. The study took place in the UK from August to November, 2009, using 36 Holstein dairy cows. The study had three 26 day experimental periods, with 12 cows studied in each. Within each period a cross over design was used with the cows swapping between a 'Control' (n=6) and 'Treatment' group (n=6) halfway through the study period. Control cows had access to TMR indoors only, whereas treatment cows had access to TMR indoors and at pasture via 'Calan' gates. As a group, the cows were given the choice of going out to pasture (1.5 ha) or into a cubicle house following AM and PM milking. They were then free to move between the two until the next milking. A video camera was used to record time spent in each location. The cows spent 71.1% (±1.82%) of their time at pasture when given the choice. One sample t-tests showed this was different from 0% (t=39.10; $P<0.001$), 50% (t=11.60; $P<0.001$) and 100% (t=-15.89; $P<0.001$). ANOVA of the percentage time spent at pasture (after logit transformation) revealed there was no interaction between treatment and study period ($F_{2,33}=0.1$, $P=0.889$). Cows spent less time at pasture as the season progressed (average of 86.7%, 68.3% and 58.3% for periods 1, 2 and 3; $F_{2,33}=44.2$, $P<0.001$) but treatment had no effect ($F_{1,33}=1.1$, $P=0.298$). The results indicated that cows had a partial preference for pasture, which was not influenced by TMR being provided at pasture, although the cows spent less time at pasture as the season progressed.

Both pen size and shape affect locomotor play in dairy calves

Mintline, Erin M.[1], De Passillé, Anne Marie B.[2], Rushen, Jeffrey P.[2] and Tucker, Cassandra B.[1], [1]University of California, Davis, Animal Science, One Shields Ave, Davis, California 95616, USA, [2]Agriculture and Agri-Food Canada, 6947, 7 Highway, P.O. Box 1000, Agassiz, BC V0M 1A0, Canada; emmintline@ucdavis.edu

Play behavior is influenced by environmental factors including housing and management. For example, the amount of space provided may affect a young animal's expression of play. Pen size is often described in terms of area per animal, but the shape of the pen may also influence play behavior. We evaluated the effect of pen size and shape on locomotor play in six-week-old Holstein calves (n=18). Calves were housed in groups of four or ten and each calf was tested individually for 15 minutes in four pens of different size and shape. Calves were visually isolated from other calves during testing. Treatments were presented in a 2×2 design where both area (large: $60m^2$, small: $30m^2$) and shape (long: length x width = 30×2 or 15×2 m, wide = 15×4 or 7.5×4 m) were varied. Duration of running and the number of bucks, jumps, and kicks were recorded continuously. A 2×2 repeated measures ANOVA with ranked values was used for analysis. Calves spent more than twice as much time running in larger (19 vs. 8s; SE=2.6 $P=0.03$) and longer pens (18 vs. 9s; SE=2.6, $P=0.05$). This is likely due to the greater distance available for uninterrupted running in these pens. Calves tended to jump more often in smaller pens (frequency= 3.3 vs. 2.4; SE=0.4, $P=0.07$), possibly because less space is required to perform this behavior. No differences were found for bucks and kicks nor was there an interaction between size and shape for any variable. There were, however, large differences between calves in the amount of play they performed. The size and shape of the pen affects the amount of locomotor play behavior shown by dairy calves. When designing pens for behavioral research or long-term housing, both shape and size need to be considered, rather than relying exclusively on measures of area.

Lying down duration in German Holstein dairy cows housed in cubicles in relation to body size and cubicle characteristics

Plesch, Gudrun and Knierim, Ute, University of Kassel, Department of Farm Animal Behaviour and Husbandry, Nordbahnhofstr. 1a, 37213 Witzenhausen, Germany; plesch@uni-kassel.de

Resting comfort and thereby cow welfare is often impaired in cubicle systems due to inadequate cubicle design and dimensions. Duration of lying down has been suggested as one measure of resting comfort. It was the aim of this exploratory study to determine which cubicle characteristics may affect this measure when taking cow body dimensions into account. Cubicle measures, diagonal body length and withers height of 80-100% of the cows depending on herd size (on average 60 cows) were recorded on 24 farms with very low within-farm variation in cubicle characteristics. 9 farms had deep bedded cubicles, 11 raised and 4 raised cubicles with bedding retainers. Durations of 30 lying down events per herd (mostly of different cows) were directly measured from carpal joint bending to completion of lying down. Compliance with recommendations for cubicle dimensions was calculated based on the largest 25% animals of the herd. Stepwise regression (SAS, proc reg stepwise) included 19 independent variables and 'duration of lying down' as dependent variable. Due to the exploratory character of the analysis, the limit for model entry was set at $P=0.5$, and 0.2 for removal. Lying down took on average 5.81 sec (range 4.96-7.64 sec). The final model explained 55% of the total variance (F=5.76, $P=0.0033$). Lying down duration was significantly shorter in deep bedded cubicles (F=11.48, $P=0.0026$). Further explanatory factors in the model were neck-rail height (F=2.75, $P=0.1121$), deep bedding or comfort mattresses versus concrete floor or rubber mats (F=2.07, $P=0.1658$) and clearance height of side partitions (F=3.13, $P=0.0930$). On average recommendations on neck-rail height were fulfilled 89% (range 77-106%) and average partition height was 54 cm (range 44-94 cm). While certain cubicle characteristics relating to cubicle floor quality and freedom of movement potentially affect lying down duration in expected directions, a considerable proportion of variance remains unexplained.

Behaviour of different dairy cattle genotypes in pasture-based production systems

Keckeis, Karin[1,2], Thomet, Peter[1], Troxler, Josef[2] and Winckler, Christoph[3], [1]Swiss College of Agriculture, Länggasse 85, 3052 Zollikofen, Switzerland, [2]Institute of Animal Husbandry and Welfare, University of Veterinary Medicine, Veterinärplatz 1, 1210 Vienna, Austria, [3]Division of Livestock Sciences, University of Natural Resources and Applied Life Sciences, Gregor-Mendel-Str. 33, 1180 Vienna, Austria; karin.keckeis@bfh.ch

Pasture-based dairy milk production as prevalent in New Zealand has been considered a cost-effective low-input production system in Switzerland. Swiss dairy cow genotypes are currently compared to New Zealand Holstein Friesian (NZL) cows in terms of their potential for pasture-based milk production. As part of this project, we studied behavioural time budgets and respiration rates (RR) on pasture during summer as indicators for heat stress. For this purpose, in total 13 Swiss RHxSi (CH) and 13 NZL cows served as focal animals on 3 farms (average herd size 40 cows). Genotype groups were frequency matched according to lactation stage and age to account for potential confounding effects. From July to mid-September on each farm focal animals were observed during three 3-day periods. Grazing, standing, lying, ruminating, clustering behaviour and RR per 30sec were recorded by direct scan sampling between morning and evening milkings. Position sensors attached to the right front leg were used to record lying times (24h) with simultaneous monitoring of environmental parameters by portable weather stations. Scan sampling and logger recordings were summarised on a daily basis and analysed using LMMs. Genotypes differed significantly in grazing and standing behaviour with CH cows spending more time grazing (48.9 vs. 43.8%, $P<0.01$) and NZL cows showing a higher proportion of standing (36.9 vs. 31.1%, $P<0.05$) though clustering was not affected. Moreover, NZL cows showed a lower daily lying time (7.8 vs. 8.3h, $P<0.05$) than CH cows, whereas there was no difference in lying and ruminating behaviour between milkings. NZL cows consistently had higher RR (e.g. 66 vs. 46bpm between 10:30 and 14:00, $P<0.001$) than CH cows. Particularly results on RR and daily lying time suggest that NZL cows might be more strained during summer pasture conditions in Switzerland. Assessment of overall welfare should include further physiological, behavioural and performance data.

Milk yield affects time budget in dairy cows

Norring, Marianna[1], Valros, Anna[1] and Munksgaard, Lene[2], [1]Research Centre for Animal Welfare, Faculty of Veterinary Medicine, P.O. Box 57, 00014 University of Helsinki, Finland, [2]Faculty of Agricultural Sciences, Aarhus University, P.O. Box 50, 8830 Tjele, Denmark; marianna.norring@helsinki.fi

The daily milk yield of dairy cows has increased considerably during past decades. However, the effect of the increased yield on cow behavior is not yet clarified. We investigated the effect of milk yield on time budget and resting using 32 cows of parities 1-7 all in the 8th week of lactation, while kept in tie-stalls. Behavior (lying, eating, ruminating, lying inactive not ruminating, lying neck muscles relaxed) and milk yield were measured for 2 days. The data was analyzed per 24 h periods, and per 4-hour periods before and after milking. The effects of milk yield and parity on the different behavioural categories was analyzed with mixed models. With increasing yield the cows spend more time eating and ruminating while standing and less time lying ($P<0.05$). Multiparous cows ruminated more than primiparous cows (435 ± 17 vs. 372 ± 18 min/d) but there were no difference in the eating (214 ± 11 vs. 227 ± 11) or lying time (701 ± 45 vs. 617 ± 47). In the morning before milking and in the evening after milking higher yielding cows were lying less compared to lower yielding cows ($P<0.05$). Latency to sleep (lie inactive not ruminating or neck relaxed) after lying down was shorter in high yielders compared to lower yielders ($P<0.05$). Cows with high milk yield needed to eat longer and thus their lying time was shorter and they fell to sleep sooner with a lying bout. The degree of milk yield affects the behavior and it is possible that high milk production can make dairy cows susceptible to time constraint.

Interactions between herd keeping stallions in a big enclosure

Sigurjónsdóttir, Hrefna[1], Thórhallsdóttir, Anna Gudrun[2], Hafthórsdóttir, Helga[2] and Granquist, Sandra Magdalena[3], [1]University of Iceland, Department of Education, Stakkahlid, 105 Reykjavík, Iceland, [2]Agricultural University of Iceland, Hvanneyri, 311 Borgarnes, Iceland, [3]Institute of Freshwater Fisheries and the Icelandic Seal Center, Hvammstangi, 530 Hvammstangi, Iceland; hrefnas@hi.is

One herd with 4 stallions and their families (bands) was studied in Iceland in May 2007 in a large pasture (215 ha). Bands sizes were 12, 20, 31 and 32. No bachelor group was present. The purpose was to map home ranges and study interactions between stallions. Each band was observed for 80 hours during daylight hours in 5 hour shifts employing the method all occurrences of some behavior. Behavior classes used were: Approach, fighting, aggressive bites, parallel walk, posturing, foreleg strike, aggressive chase and marking. Also, interactions with one intruding stallion which was released with 10 mares into the enclosure were recorded. To estimate boundaries of home ranges location of observed bands was marked every hour (ArcGIS used for analyzing). Usually the stallions kept their bands along the only water ditch such that the middle stallions had two neighbours, the others one. Home ranges overlapped, but bands never mixed. Generally, interactions between resident stallions were not aggressive - they assessed each other without much contact. They did not try to steal mares from each other, in 5 cases they drove young mares in their group away. Most interactions occurred between neighboring stallions ($\chi^2 = 88.7$ (d.f.= 1), $P<0.001$). The intruding stallion lost most of his group members and after 8 days he was removed. Encounters between him and the resident stallions were more aggressive (frequency of fights, aggessive chases and bites:$\chi^2 = 161.7$ (d.f.= 1), $P<0.001$) and more frequent than between the others ($\chi^2 =11.61$ (d.f.= 1), $P<0.01$). This study shows how access to water shapes home ranges. Because stallions are protective of their bands it will cause stress if access to water is limited. The study also shows that introduction of new groups into a stable herd will upset the horses and increase aggression.

Graded leadership of dominant beef cows: stronger in space than in time?

Sarova, Radka, Spinka, Marek, Panama, Jose A. and Simecek, Petr, Insitute of Animal Science, Pratelstvi 815, 104 00 Praha, Czech Republic; sarova.radka@vuzv.cz

In herd of ungulates, synchronisation in time and coordination in space are prominent phenomena that require agreement about activity switches and about movement trajectories. Do herd members contribute equally to these decisions or are they controlled by a leader/ leaders, e.g., by dominant animals? The aim of our study was to assess whether more dominant cows exert more influence on activity changes and movements in a beef cattle herd on a pasture than subordinate animals. For all adult cows in the herd, (n=15), positions were recorded with GPS collars every minute, exact times of all switches between activity and lying were noted visually and behaviour was scanned in a 5 minute interval between 5am-9pm during 15 days. Based on eight movement variables such as speed, distance between individuals, herd shape and movement alignment, cluster analysis divided data points into three herd behaviour types: resting, foraging and short-distance travelling. More dominant cows did not induce more changes from activity to resting (r_s = -0.16, n = 14, $P>0.05$) or from resting to activity (r_s = -0.38, n = 14, $P> 0.05$) than the more subordinate cows. However, more dominant animals were found more in the front of the herd during both foraging (r=0.63, n=15, $P<0.05$) and short-distance travelling (r=0.66, n=15, $P<0.05$). More dominant cows also covered shorter distances (r=-0.54, n=15, $P<0.05$) during foraging and had straighter travel trajectories (r=0.67, n=15, $P<0.05$) and higher alignment with the herd (r=0.67, n=15, $P<0.05$) and with their neighbours (r=0.67, n=15, $P<0.05$) during short-distance travelling. Nevertheless, at any given minute, the effect of dominance on movement variables was limited, with determination coefficients attaining maximally 9%. In conclusion, beef cows in our study exerted influence on herd movements, especially during short-distance travelling, that was graded along their dominance status, but such effect was not found for synchronisation in time.

Effect of tree cover on ranging behaviour of free range hens

Cooper, Jonathan and Hodges, Holly, University of Lincoln, Animal Behaviour, Cognition and Welfare Research Group, Biological Sciences, Riseholme Park, Lincoln, Ln2 2LG, United Kingdom; jcooper@lincoln.ac.uk

Free-range layer legislation requires large outdoor ranges, but hens rarely make effective use of this area, especially in large flocks. We investigated ranging behaviour of 20 commercial free-ranging flocks (over 5,000 hens per flock) which varied in degree of natural tree cover between 10m and 50m from sheds. Four transects (5m wide and 100m long) were laid out at right angles to the sides of shed and for each 5m quadrat (i.e. 25m^2) we estimated % tree cover and counted number of hens four times per flock from approx. 12:00 h to 16:00 h. Seven flocks had no tree cover between 10m and 50m from shed (No Cover), 5 flocks had less than 1m^2 of tree cover (Low), 4 flocks had between 1m^2 and 10m^2 (Medium), and 4 flocks had an average of over 10m^2 of cover per 25m^2 quadrat (High). The highest hen densities were found near the shed; 8.1±1.3 hens per quadrat between 0m and 10m from shed. There was no difference between No, Low, Medium and High cover flocks on near-shed density (One way ANOVA; F3,16 = 0.36, ns). In the planted zone between 10-50m from shed there was a strong influence of canopy cover (One way ANOVA; F3,16 = 8.91, P<0.001) with only 1.0±0.2 hens per 25m^2 with no cover rising to over 3.3±0.4 hens when there was more than 10m^2 of cover per quadrat. Beyond 50m from the shed there was less than 0.2 hens per quadrat and there was no effect on canopy cover (One way ANOVA; F3,16 = 1.58, ns). These findings indicate that providing natural cover in the form of trees can increase number of hens using outdoor range in free range flocks, especially where canopy cover is high.

Predicting tail biting outbreaks among growing-finishing pigs under commercial conditions

Wallenbeck, Anna[1], Larsen, Anne[1], Holmgren, Nils[2] and Keeling, Linda J.[1], [1]Swedish University of Agricultural Sciences, Department of Animal Environment and Health, P.O. Box 7068, SE-750 07 Uppsala, Sweden, [2]Swedish Animal Health Service, Swedish Animal Health Service, SE-532 89 Skara, Sweden; Anna.Wallenbeck@hmh.slu.se

Predicting tail biting (TB) could help the stockperson prevent severe outbreaks and thus reduce pigs' suffering as well as production losses. The present study aimed to identify behavioural indicators of TB outbreaks. We registered pig behaviour in 9 pens on a commercial farm. Ultimately 3 TB pens (TBPs) were matched with 3 neighbouring control pens (CPs) without TB. Pig behaviour before (range -30 to -2, mean -17 days before bloody tails were observed in the pen (start of TB outbreak)) and during TB outbreaks (range 0 to 13, mean 5 days after start of TB outbreak) was compared between TBPs and CPs using linear mixed models correcting for expected changes in behaviour over time. Pigs in TBPs tended to spend more time rooting before a TB outbreak ($P=0.091$) and left the feed trough before the feed was finished more often ($P=0.017$) than pigs in CPs. Regarding social interactions before the TB outbreak, actors in TBPs delivered more head knocks ($P<0.001$) and receivers walked away ($P=0.005$) and vocalized ($P=0.054$) more than pigs in CPs. However, the decrease in the frequency of head knocks (before vs. during TB outbreak) was greater in TBPs than it was in CPs ($P=0.006$). As expected, frequency of biting (towards tail, ears or body) increased in TBPs before vs. during TB outbreak, while it decreased in CPs ($P=0.006$). The proportion of pigs with non-curled tails in TBPs increased significantly from 5% before to 28% during the TB outbreak ($P<0.001$). In conclusion, our results suggest that there are early behavioural indicators of TB (1-4 weeks before an outbreak). The data indicate that higher levels of foraging behaviour, away from the feed trough, and increased hostile social interactions, more vocalisation and avoidance, may distinguish a potential future TBP, whereas non-curled or bloody tails indicate that TB has started.

The effect of flock size and paddock complexity on following behaviour in Merino sheep

Taylor, Donnalee[1], Price, Ian[2], Brown, Wendy[1] and Hinch, Geoff[1], [1]UNE, Animal Science, Armidale, NSW, 2351, Australia, [2]UNE, Psychology, Armidale, NSW, 2351, Australia; dtaylor2@une.edu.au

Little is known about the impact of flock size on the social structures (leadership and sub-grouping) of the highly gregarious Merino sheep. As part of a larger experiment examining the capacity to manipulate flocks using 3 leaders trained to approach a stimulus for a lupin grain food reward, the present experiment examined the impact of flock size on responsiveness to leader-initiated naïve (non-trained) group movement and sub-grouping formation in small paddocks (2 ha). Two groups SM (Small Mob, n=18, 3 trained + 15 naive) and LM (Large Mob, n=48, 3 trained + 45 naive) were tested during Morning Grazing and Afternoon Grazing in 3 open paddocks (OPs) and 3 complex paddocks (CPs). In all 6 tests 100% of the SM followed leader initiated movement approaching within 6m of the stimulus in OPs and CPs. The number of LM group members following changed significantly with more sheep following in the OPs than the CPs (Chi-square(df3)=6.39, P=0.012). The gregarious nature of sheep, their social cohesiveness and allelomimetic behaviour seemed to facilitate group movement. Passive recruitment by leaders and associated following behaviour of naive sheep was observed consistently in the OPs but in the LM group complexity did reduce the influence of leaders. Overall sub-grouping did not change (Chi-square (df3)=0.26, P=0.97) with group size and did not significantly alter response to leaders in the CPs or OPs. Sub-grouping in both the SM and LM CPs increased by three additional sub-groups in the afternoon compared with the morning. This may be a reflection of high pasture availability for morning grazing subsequently reducing emphasis on food gathering in the afternoon i.e.: social interactions rather than hunger needs became predominant. It seems that in small complex paddocks sub-grouping may be related more to level of social activity than to group size or paddock complexity per se.

Social behaviour during winter outdoor exercise of cows of the Hérens breed kept in tie-stalls

Castro, Isabelle M.L., Gygax, Lorenz, Wechsler, Beat and Hauser, Rudolf, Center for Proper Housing of Ruminants and Pigs, Federal Veterinary Office, Agroscope Reckenholz-Tänikon Research Station ART, 8356 Ettenhausen, Switzerland; isabelle.castro@art.admin.ch

Cows of the Hérens breed, selected for both milk and beef production, are highly motivated to engage in dominance interactions and famous for the cow fights traditionally organised by Swiss breeders. However, this characteristic may result in excessive aggressive behaviour when cows kept in tie-stalls meet during winter outdoor exercise. In this study, we varied the length of the interval between days with winter outdoor exercise (daily, every 3rd, 4th or 5th day) in all herds of 6 Swiss working farms including 51 cows of the Hérens breed (3, 4, 5, 7, 13 and 19 cows, respectively). The behaviour during outdoor exercise was observed and the appearance of new injuries that took at least 3 days to heal were noted for the head, udder and remaining body. The total duration of each exercise period (40 minutes) was divided into four blocks of 10 minutes. Agonistic and allogrooming behaviour as well as injury scores were analysed using generalized linear mixed-effects models. The frequency of agonistic behaviour increased significantly with increasing length of the interval between exercise days (mean±StdErr, daily exercise: 2.15±0.07, every 3rd day: 2.10±0.10, every 4th: 2.87±0.12, and every 5th: 3.59±0.14 interactions/10min; $F_{1,171}=64.628$, $P<0.001$) and decreased significantly with the duration of the exercise on a given day (from 3.16±0.12 for the 1st to 2.04±0.09 for the last 10 min; $F_{1,2050}=82.756$, $P<0.001$). Duration of allogrooming was not significantly influenced by neither the interval between nor the time within exercise days. Except for the head, the risk for injuries increased significantly with increasing length of the interval between exercise days (udder: OR=1.81/day, $t_{140}=3.43$, $P<0.001$; remaining body: OR=1.59/day, $t_{140}=4.29$, $P<0.001$). In conclusion, cows of the Hérens breed show pronounced agonistic behaviour after only a few days of tethering. Consequently, the length of the interval between exercise days should be as short as possible.

Effects of three methods for mixing unfamiliar horses on their potential to minimise aggressive interactions

Hartmann, Elke[1], Keeling, Linda J[1] and Rundgren, Margareta[2], [1]Swedish University of Agricultural Sciences, Animal Environment and Health, P.O. Box 7068, 750 07 Uppsala, Sweden, [2]Swedish University of Agricultural Sciences, Animal Nutrition and Management, P.O. Box 7024, 750 07 Uppsala, Sweden; Elke.Hartmann@hmh.slu.se

This study compared three mixing methods to assess their potential in minimising aggression between horses: (1) one unfamiliar horse met one resident horse (P, 10 min, 15 tests), (2) one unfamiliar horse met two resident horses (PP, 10 min, 15 tests), (3) one unfamiliar horse was pre-exposed to one resident horse in neighbouring boxes, containing an opening to allow horses to put their head through (B, 5 min, 16 tests), afterwards horses met in the paddock (BP, 10 min, 16 tests). Horses were 16 Standardbred mares (6-18 years) and the order of mixing methods was balanced. Social interactions were observed and behaviour recorded as frequency per test. There were no differences in total aggression between methods and none of the tests needed to be terminated. Unfamiliar horses did not receive more aggression from resident horses in PP than P (Mean±SD, PP=5.08±3.14, P=3.28±2.43, P=0.22), including those aggressive behaviours where physical contact is attempted (e.g. kick, strike). However, 'attack' was more frequent in PP than P (Median (IQR), PP= 2 (0, 5), P=0 (0, 1), P=0.04) and 'flee' was more frequent (PP=6 (4, 8), P=1 (0, 6), P=0.02). Pre-exposure in boxes did not reduce aggression in BP compared to P (Median (IQR), BP=7 (4.25, 11.75), P=6 (2, 16), P=0.77) but horses exchanged more non-aggressive (Median (IQR), 2 (0.25, 4)) than aggressive (0 (0, 1), P=0.013) and mixed (0 (0, 1), P=0.02) interactions through the opening in B. Results suggest that mixing an unfamiliar horse with two resident horses, instead of one by one, may be preferable. Although the mixing may be more stressful, because the aggression is more intense, total aggression received by the unfamiliar horse would potentially be less. We further suggest, although not specifically supported by current results, to allow horses to interact in a safe box environment prior to mixing.

Use of outdoor range in large groups of laying hens

Gebhardt-Henrich, Sabine G. and Fröhlich, Ernst, Centre for proper housing: poultry and rabbits, Burgerweg 22, CH-3052 Zollikofen, Switzerland; sabine.gebhardt@bvet.admin.ch

Under commercial conditions laying hens are kept in unnaturally large groups. Several studies have shown a negative correlation between group size and use of an outdoor range. However, previous studies did not distinguish individual hens. In our study of eight flocks (2,000 - 18,000 hens) individual ranging behavior was registered during 17 - 29 days. Besides a hen house with aviaries hens had a covered and an open outdoor range. Hens had to cross the covered part to get to the open range. 5 - 10% of the hens were fitted with RFID tags and flat antennae were placed on the floor in front of and behind each pophole. Reliability of recording varied among days and was between 11 and 100% (mean = 78 ± 0.02). Repeated measures analyses were performed. 48 - 90% of the tagged hens used the open outdoor range, 76 - 99% used the covered outdoor range at least once during the observation period. The percentage of hens using the outdoor range (x = 68.00% ± 14.99) and the time spent outside (x = 53.66 ± 28.53 min. on a day) were not related to the size of the flock (percentage of hens: r^2 = 0.2, t = 9.12, NS, time: $F_{2,1852}$ = 1.35, NS). The number of trips (x = 4.45 ± 1.76) to the outdoor range depended on the size of the flock ($F_{2,1852}$ = 4.83, P= 0.008), hens of medium flocks went outside significantly more often than hens of small or large flocks. Only about 20% of the hens used the open range every day but most hens went into the covered part every day. This study shows that the majority of hens in large flocks use outdoor ranges even when only few of them can be seen outside at any one time.

Differences in hypothalamic gene expression in feather pecking chickens

Brunberg, Emma[1], Jensen, Per[2], Isaksson, Anders[3] and Keeling, Linda[1], [1]Swedish University of Agricultural Sciences, Department of Animal Environment and Health, P.O. Box 7068, SE-750 07 Uppsala, Sweden, [2]Linköping University, Department of Biology, SE-581 83 Linköping, Sweden, [3]Uppsala University, Department of Medical Sciences, Uppsala Array Platform, SE-751 85 Uppsala, Sweden; emma.brunberg@hmh.slu.se

Feather pecking (FP) is suggested to be redirected pecking behaviour related to birds' foraging motivation. Several studies have also suggested that the genetics of a bird can predispose it to developing the behaviour. This study aimed to explore the biological mechanisms underlying FP by using gene expression microarrays. Cages of birds (Lohmann Selected Leghorns) on a commercial egg producing farm were pre-selected based on the plumage condition of the birds. The FP behaviour of birds in these cages (n=209) was then observed. Based on the number of performed or received severe feather pecks, 31 birds in 10 matched groups were categorized as peckers (n=11), receivers (n=10) or neutral birds (n=10). The Affymetrix GeneChip® Chicken Genome Array was used for profiling gene expression in the hypothalamus from these birds. An empirical Bayes moderated t-test was used with a correction for multiple testing to search for differences between the groups. Peckers performed more but received fewer ($P<0.05$) pecks per minute (1.16 ± 0.24, mean ± SE; 0.09 ± 0.07) than receivers (0.01 ± 0.003; 0.26 ± 0.05). The neutral birds did not perform or receive any pecks. The expression of 16 transcripts differed significantly (adjusted $P<0.5$) or showed a tendency (adjusted $P<0.1$) to differ in one or more of the three possible comparisons of these categories of birds. In the relatively short gene list, possible links to insulin resistance (hence feeding behaviour), chronic intestinal inflammation and obsessive compulsive disorder (OCD) could be distinguished. These results may support the earlier proposed hypothesis that the behaviour is an abnormal foraging behaviour and that it might serve as an animal model for OCD. In conclusion, the gene list presented in this study may after confirmation be of use for further research on the biological mechanisms behind FP.

The pigmentation regulating PMEL17 gene has pleiotropic effects on behavior in chickens

Karlsson, Anna-Carin[1], Mormede, Pierre[2], Kerje, Susanne[3] and Jensen, Per[1], [1]Linköping University, IFM, Biology, Linköping University, 58381 Linköping, Sweden, [2]Université Victor Segalen Bordeaux 2, 2PsyNuGen-Neurogenetics and Stress, INRA UMR1286, CNRS UMR5226, 146 rue Léo-Saignat, 33076 Bordeaux Cedex, France, [3]Uppsala University, Department of Medical Biochemistry and Microbiology, Uppsala biomedicinska centrum BMC, Husarg. 3, 75123 Uppsala, Sweden; ankar@ifm.liu.se

Earlier studies have shown that the PMEL17 gene not only affects plumage pigmentation, but also has pleiotropic effects on behavior in chicken. For example, a mutation in the gene protects the carrier against being feather pecked, and appears to affect social and explorative behavior. However, previous studies have not been able to tear apart the direct effects of genotype on behavior from indirect effects caused by different negative social experiences, for example from being feather pecked. In this study chickens from an advanced White Leghorn x red jungle fowl intercross homozygous for the Dominant white (9 females; 6 males) or wild-type (14 females; 16 males) allele of PMEL17, were subjected to a broad phenotyping in order to detect consistent behavioral differences between genotypes. To exclude negative social experience the chickens were individually housed without physical contact from the day of hatching, and tested for social, aggressive, fear and exploratory behaviors between 14 and 294 days of age. Further, costicosterone and testosterone levels were assessed. In a principal component analysis, 53.2% of the behavior variation was explained by two factors. Factor one was an activity and social factor, and there was a significant effect ($F1,30 = 8.68$, $P<0.01$) of genotype on the factor scores. On factor two, related to aggressive behavior, there were significant effects of genotype ($F1,30 = 6.39$ $P<0.05$), sex ($F1,30 = 6.09$ $P<0.05$) and their interaction ($F1,30 = 6.79$ $P<0.01$). There were no genotype effects on hormone levels or any other measured non-behavioral phenotypes. Hence, differences in behavior between PMEL17 genotypes remained when negative social experiences were excluded, indicating a direct pleiotropic effect of the gene on behavior.

Cuteness and playfulness in dogs, are they related?

Forkman, Björn, University of Copenhagen, Department of Large Animal Sciences, Groennegaardsvej 8, 1870 Frederiksberg C, Denmark; bjf@life.ku.dk

In an influential article was published on 'behavioural paedomorphism' i.e. that there is a relation between the aggressive behaviour and the physical similarity of the breed to the wolf. The aim of the current study is to investigate the behavioural paedomorphism hypothesis using a different motivation; playfulness (considered to be a characteristic of young animals). In the first study the results from the Swedish Dog Mentality Assessment (DMA) for the breeds included in Goodwin et al's study were compared. In the present study three of the DMA-measures were used: Interest in play, Grabbing a rag when it has been thrown, and tug of war. The behaviour was scored on a 5-point scale, from low to high intensity. The breeds were categorised as least wolf-like: Cavalier King Charles (n=50), Norfolk Terrier (n=24), French Bulldog (n=39), Cocker Spaniel (n=156); slightly wolf-like: Shetland Sheepdog (n=179), Large Munsterlander (n=3), Golden Retriever (n=2592) and Labrador Retriever (n=1616); most wolf-like: German Shepherd (n=13371) and Siberian Husky (n=9). If the results of all breeds are included the least wolf-like breeds had an average of 3.1 (Interest), 2.6 (Grab) and 2.1 (Tug), with the slightly wolf-like breeds had 3.7, 3.2 and 3.0, and the most wolf-like breeds had 3.4, 3.1 and 2.8, i.e. almost the reverse of what was predicted. In the second study nose length/face height was used as an indicator of cuteness i.e. puppy-likeness or cuteness (sensu Lorenz). Seventeen breeds were included in this study, all with more than 200 individuals tested in the DMA. There was no significant correlation between the cuteness score and the Interest in play (Rs=0.22, P=0.4), Grabbing a rag (Rs=0.35, P=0.17), or tug of war (Rs=0.19, P=0.47). The hypothesis that less wolf-like dog breeds show more puppy-like behaviour is not supported in the current studies.

Heritabilities of birth assistance and neonatal lamb traits in Suffolk sheep

Matheson, Stephanie, Macfarlane, Jennifer, Dwyer, Cathy and Bünger, Lutz, Scottish Agricultural College, Sustainable Livestock Systems, West Mains Road, Edinburgh, EH9 3JG, Scotland, United Kingdom; Stephanie.Matheson@sac.ac.uk

Two causes of lamb mortality are dystocia and low vigour lambs. Both of these problems require high levels of human intervention to ensure lamb survival. Therefore, reducing the occurrence of these problems can improve lamb welfare while reducing labour input. This study aimed to explore the feasibility of selection for improved welfare of neonatal Suffolk lambs through selection of parents with superior vigour characteristics by estimation of genetic parameters. Neonatal data was collected in 2008 and 2009 from 290 pedigreed flocks from members of the Suffolk Sheep Society. Each lamb was scored on: birth assistance (BA; 17,360 records), where 0- unassisted within 30 min, 1- unassisted within 30-60 min, 2- minor assistance required, 3- major assistance required, 4- veterinary assistance required; lamb vigour (LV, at 5 minutes of age; 17,387 records), where 0- extremely active lamb, has been standing on all 4 feet, 1- very active lamb, standing on back legs, 2- active lamb, on chest, holding head up, 3- weak lamb, lying flat, able to hold head up, 4- very weak lamb, unable to lift head; sucking assistance (SA; 17,227 records), where 0 - no assistance, sucking within 1 hour, 1- no assistance, sucking within 2 hours, 2- given assistance once or twice in first 24 hours, 3- given sucking assistance more than twice, needing help between 1-3 days, 4- still needing help when more than 3 days old. Heritabilities were estimated by fitting univariate individual animal models using ASREML. Significant effects were included in the linear mixed model. Heritabilites (h2) for all neonatal traits were moderate; BA = 0.27 (s.e. 0.034); LV = 0.34 (s.e. 0.037); and, SA = 0.19 (s.e. 0.031). These results indicate that BA, LV and SA could be included in breeding programmes to improve welfare and reduce labour input in Suffolk flocks.

Transgenerational epigenetic inheritance of behavioural traits in the chicken (*Gallus gallus*)

Nätt, Daniel and Jensen, Per, Linköpings University, IFM Biology, division of Zoology, Linköpings University, 58183 Linköping, Sweden; danis@ifm.liu.se

Genetic selection is considered to act on base-pair mutations, but acquired epigenetic changes have also potential to be targets of selection. Previously, we showed that chickens may transmit stress induced behavioural traits and associated brain gene expression to its offspring. Here we further explores inheritance of behaviour and gene expression by comparing: Breed (Red Junglefowl [RJF] vs. domesticated White Leghorn, [WL]), with Family (a high fearful [hi] vs. low fearful [lo] line selected by standardized fear test scores). We hypothesized that family inheritance will be high due to epigenetic factors. Offspring [Off] (nRJF-hi=17; nRJF-lo=19; nWL-hi=20; nWL-lo=17) were hatched from a breeding pair [Par] of each breed and family (nRJF-hi=1+1; nRJF-lo=1+1; nWL-hi=1+1; nWL-lo=1+1). Hatch weight differences (g) between the families were inherited within breeds (means±SEM: Off-WL-hi= 49.9±0.7 vs. Off-WL-lo=44.4±0.5, F1:33=41.0, $P<0.001$; Off-RJF-hi=32.4±0.4 vs. Off-RJF-lo=27.7±0.3, F1:32=94.3, $P<0.001$), while fear level was only inherited in domesticated offspring (fearfulness score: Off-WL-hi= 488.5±19.7 vs. Off-WL-lo=408.6±34.5, F1:33=4.4, $P<0.05$). Hypothalamic RNA from each breed, family and sex was hybridized to gene expression microarrays (nPar=8, nOff=8). As expected, gene expression differences were more common between breeds than between families within breed. Of the genes differentially expressed between breeds in the adult parents (FDR adjusted P-value<0.05), 86% re-emerged as significant in the same comparison in three the weeks old offspring, indicating high age independency in the effect of domestication. These overlapping genes also showed a maternal effect (mean r parent to offspring, females=0.962±0.002 vs. males=0.942±0.006, $P<0.01$). Using fold ranks [FR], 40% of the 1000-top-ranked genes of the parents were also found on the corresponding top-list of the offspring. Interestingly, hi and lo families within breed had similar overlap (WL=32%, RJF=35%). One gene (sorcin) was up-regulated in hi vs. lo in all comparisons (Par-RJF, FR=1; Par-WL, FR=1; Off-RJF, FR=1; Off-WL, FR=7 out of 23,169 transcripts). We conclude that the transgenerational effect of gene expression was high in both breeds and families, indicating that phenotypic differences within a breed may be mediated by heritable epigenetic factors.

Social bonding and sociability are affected by the level of social contact in dairy calves

Duve, Linda Rosager and Jensen, Margit Bak, University of Aarhus, Faculty of Agricultural Sciences, Blichers allé 20, P.O. Box 50, 8830 Tjele, Denmark; linda.duve@agrsci.dk

Dairy calves are often raised individually with limited opportunity for social contact. They are, however, highly motivated for full social contact to peers, and the present study investigates the effects of the level of social contact. Twenty-four pairs of newborn calves were reared for six weeks with either limited social contact through bars (L-calves); limited contact for three weeks followed by full contact for three weeks (L/F-calves) or full social contact for six weeks (F-calves). From day 42 calves were housed in groups with 2 calves from each treatment. Preference for the companion compared to an unfamiliar calf was evaluated in a triangular arena on day 35, and the latency to approach and duration of social contact were recorded. In the group pens preference and sociability were assessed by looking at nearest neighbour during rest. Data were analysed using a general linear mixed model (SAS). During the preferences test the companion was approached before the unfamiliar calf in 87.5% of the cases for F-calves, 37.5% for L/F-calves and 14.3% for L-calves (Fishers exact test, $P=0.007$), and the F- and L/F-calves spent significantly more time with the stimuli calves than the L-calves ($F_{2,13}=7.60$; $P<0.01$). But within treatment there was no difference in time spent with each stimuli calf ($F_{1,39}=0.78$; $P=0.47$). In the groups, no significant difference was found between treatments in time spent with the companion ($F_{2,69}=0.29$; $P=0.75$). The calves on all treatments spent significantly more time with the companion on day 43 compared to day 49 ($F_{1,69}=25.47$; $P<0.0001$). The F-calves spent significantly more time in body contact with their neighbours than the L-calves ($F_{2,65}=5.15$; $P<0.01$). Thus, the level of contact to peers apparently affects the sociability of young dairy calves, but even calves with limited opportunity for social contact have a preference for the companion when grouped with other calves.

Space requirements for group housed pigs at high ambient temperatures

Spoolder, Hans A.M.[1], Aarnink, André J.A.[1] and Edwards, Sandra A.[2], [1]Wageningen UR Livestock Research, P.O. Box 65, 8200 AB Lelystad, Netherlands, [2]University of Newcastle, Dept. of Agriculture, King George VI, NE17RU Newcastle, United Kingdom; hans.spoolder@wur.nl

Housing systems which meet the behavioural needs of finishing pigs offer sufficient static space (occupied by the body of the pig), activity space (for e.g. feeding and dunging) and interaction space (for appropriate social behaviour). Estimates for static space have been presented for thermoneutral conditions, but are expected to increase substantially as temperature increases. The present paper models the relationship between ambient temperatures above the comfort zone, and thermoregulatory lying behaviour in finishing pigs. A literature review was conducted to collect information on three aspects of lying behaviour: lying frequency, posture (lateral, semi lateral or ventral lying) and level of space sharing (huddling) in response to increasing temperatures. Estimates of the effect of posture on floor occupation were obtained and presented as 'k-values' (k = floor area occupied (m^2)/body weight$^{2/3}$ (kg)) to correct for the effect of pig size. Subsequently, relationships between lying frequencies / posture changes and ambient temperature were included in the model. Model simulations suggest a limited effect of increased (lateral) lying on lying space requirements (e.g. k=0.0345 at 21 °C vs k=0.0364 at 35 °C; equivalent to 0.793 vs. 0.836 m^2 for a 110 kg pig). However, in combination with complete spatial separation between animals at high temperatures (i.e. no space sharing) the k-value increases to 0.0404 at 35 °C (or 0.929 m^2 lying space for a 110 kg pig). The exact relationship between the degree of space sharing and the pig's upper critical temperature (UCT) is currently being investigated. We conclude that lying space occupied by group housed pigs increases substantially with temperature, mainly due to reduced space sharing. This increase can be modelled to support the design of housing systems for finishing pigs.

Why wild-caught animals seldom stereotype

Jones, Megan Anne[1], Pillay, Neville[1] and Mason, Georgia[2], [1]University of the Witwatersrand, Animal, Plant, and Environmental Sciences, Jan Smuts Avenue, Braamfontein, Johannesburg 2050, South Africa, [2]University of Guelph, Animal and Poultry Sciences, 50 Stone Road East, Guelph NWG 2W1, Canada; megan.jones@icon.co.za

Wild-caught (WC) animals are generally less susceptible than captive-bred (CB) animals to developing stereotypic behaviour (SB). The striped mouse, Rhabdomys, is no exception: animals trapped as adults are a third less likely to develop SB than their CB offspring (16% vs. 57%; Wald χ^2_1=12.80, P=0.01). In this study, we investigated four potential explanations for the birth-origin effect: (1) forebrain development is normalised in subjects maturing in the wild; and/or (2) WC individuals adjust better to captivity than CB conspecifics, or instead (3) are precluded from performing stereotypy due to anxiety/fear or (4) apathy. We ran 30 WC and 14 CB striped mice through a series of behavioural tests to quantify anxiety and forebrain function, measured faecal corticosterone (n=31) as a physiological index of stress, and assessed home-cage behaviour. WC individuals, whether stereotypic or not, were more anxious/fearful than CB striped mice (e.g. proportion of time in dark compartment of light-dark box; $F_{1,37}$=5.87, P=0.02) and had raised levels of corticosterone ($F_{1,24}$=8.11, P=0.01). In their home-cages, WC non-stereotypic striped mice also spent the most time in their nest boxes (Wald χ^2_1=7.73; P=0.01), suggestive of apathy. Irrespective of birth-origin, stereotypic individuals showed more predictable behavioural sequencing in the four-arm maze than non-stereotypic animals (Wald χ^2_1=10.99, P<0.001) and, because the incidence of SB was markedly lower in WC striped mice so, too, were mean levels of routine-proneness. These results show that WC striped mice are less susceptible than CB individuals to develop SB because a higher proportion of animals benefit from normal forebrain function, with apathy possibly also contributing to the absence of SB in some WC individuals. More broadly, these findings indicate that whilst the absence of SB is usually taken to indicate conditions associated with good welfare, the low levels of SB in WC striped mice are an important exception.

Recurrent perseveration correlates with cage stereotypic behaviour and is insensitive to environmental enrichment in American mink, Mustela vison

Dallaire, Jamie and Mason, Georgia J, University of Guelph, Animal and Poultry Science, 50 Stone Road East, Building 70 Guelph, Ontario, N1G 2W1, Canada; jdallair@uoguelph.ca

Recurrent perseveration is the tendency to inappropriately repeat actions. It reflects depressed activity of the corticostriatal motor loop's indirect pathway, e.g. due to lesions or drugs, causing failure to inhibit repetition of already completed behaviours. Thus, recurrent perseveration is a neurocognitive correlate of stereotypic behaviour in autistic and schizophrenic humans, and in several captive species. We investigated whether this also correlates with stereotypic locomotive and upper body movements in caged mink. We tested 17 females (aged 2-4 years), pre-selected as stereotypic, in a serial two-choice guessing or 'gambling' task, in which opening one of two doors yielded a food reward, the other a punishment (time-out). The 'correct' door was randomly determined for each trial. Multiple indices of recurrent perseveration predicted stereotypic time budgets, quantified via scan sampling. Highly stereotypic animals produced more patterned, less random response sequences (all tests GLM: detrended fluctuation analysis – F=7.07, P=0.020; third-order Markov model – F=3.88, P=0.071), primarily because they more often repeated their previous responses (F=6.49, P=0.024). Consistent with disinhibition of previously performed behaviour, response latency correlated negatively with stereotypic behaviour for repetitions (F=9.50, P=0.009), but not for alternating responses (F=1.05, P=0.325). Subjects were then exposed to enriched environments, which reduced stereotypic behaviour by nearly two thirds (F=9.07, P=0.005), but did not affect recurrent perseveration (all indices – F<0.47, P>0.500). In addition to these older females, we tested 8 barren- and 6 enriched-raised young adults for recurrent perseveration, again finding no effect of housing condition (F<0.06, P>0.818) despite group differences in stereotypic behaviour (F=11.41, P=0.005). Thus, individual differences in recurrent perseveration seem to reflect normal variation, not dysfunction caused by barren captive environments, but recurrent perseveration predisposes toward stereotypic behaviour in such conditions.

Health in tail biters and bitten pigs in a matched case-control study

Munsterhjelm, Camilla[1], Brunberg, Emma[1], Palander, Pälvi[2], Sihvo, Hanna-Kaisa[3], Valros, Anna[2], Keeling, Linda[1] and Heinonen, Mari[2], [1]Swedish University of Agricultural Sciences, Department of Animal Environment and Health, P.O. Box 7084, 750 07 Uppsala, Sweden, [2]University of Helsinki, Department of Production Animal Medicine, PL 57, 00014 Helsinki, Finland, [3]University of Helsinki, Department of Basic Veterinary Sciences, PL 57, 00014 Helsinki, Finland; camilla.munsterhjelm@helsinki.fi

Health differences in tail biters and victims were investigated on a problem farm. Quartets (n=16) of age- and gender-matched individuals including a tail biter (TB, n=16), a victim (V, n=16), a control in the same pen (Ctb, n=10), and one in an adjacent pen without tail biting (Cno, n=14) were chosen by direct behavioural observation. The pigs were euthanized, autopsied and their tissues (liver, spleen, kidney, lung, heart, pancreas, adrenals, stomach, jejunum, ileum, colon, knee and elbow joint capsules, the tail, hypophysis, and sacral medulla) examined histologically. If the related samples Friedman's Two-Way ANOVA by Ranks or Cochran's Q test indicated a group effect ($P<0.1$) the variable was analyzed pair-wise within the quartets using Wilcoxon Signed Ranks or McNemar's Test for dichotomous or categorical variables, respectively. The number of received tail bites was stronger correlated with histological than with gross lesion scoring (Pearson $r=0.6$, $P<0.001$ vs. $r=0.5$, $P<0.001$), probably due to several tails with infections beneath healthy skin. The histological score (range 0-4) was larger in V than C_{no} (3.8 ± 0.1 [SEM] vs 1.4 ± 0.2, $P<0.01$), TB (2.0 ± 0.4, $P<0.01$) and C_{tb} (1.8 ± 0.5, $P<0.05$). The score was larger in TB than C_{no} ($P<0.05$). Multiple-organ changes indicating systemic infection were common. Respiratory organ infections (in 78% of the animals) were histologically scored as more chronic in V than in all other groups (Friedman's test $P<0.05$). The lesion score in the cutaneous part of the stomach was twice as large in TB and V than in C_{tb}, but the difference was non-significant. Cutaneous lesions are associated with stress. In conclusion, tail bitten pigs suffered from more severe respiratory infections than their penmates. Illness may predispose for being bitten and/ or be the consequence of long-term biting. The number of tail bites an animal is receiving may predict tail lesions better than gross tail appearance.

The risk of tail-biting in relation to level of tail-docking

Thodberg, Karen[1], Jensen, Karin H.[1] and Jørgensen, Erik[2], [1]University of Aarhus, Faculty of Agricultural Sciences, Department of Animal Health and Bioscience, Blichers Allé 20, P.O. Box 50, DK-8830 Tjele, Denmark, [2]University of Aarhus, Faculty of Agricultural Sciences, Dept. of Genetics and Biotechnology, Blichers Allé 20, P.O. Box 50, DK-8830 Tjele, Denmark; Karen.Thodberg@agrsci.dk

Tail-docking is used in many European countries as a preventive treatment to avoid tail-biting. Some studies have shown that docking is effective, whereas others are less conclusive and at present the risk of tail-biting in relation to docking level is still unclear. This study compared risk of tail-biting in pigs tail-docked leaving 2.9 cm; 5.7 cm; 7.5 cm; or the whole tail on the pigs, which is approx. ¾, ½, ¼ or no part of their tail removed. The pigs were housed in 4 production herds and results from a total of 91 pens were included in the study. In the finisher period the pigs were housed in groups of 18-25 in pens with different degrees of slatted floor. Two herds were fed dry feed ad. lib. and two herds restricted wet feed. The pigs were docked litter wise and never met pigs with other tail lengths than themselves. In the finisher section the pens were inspected weekly for tail-biting outbreaks, defined as at least one bleeding wound on the tail tip. Behaviour was registered weekly until an eventual tail-biting outbreak. The tail lengths were measured at introduction to weaner and finisher pens and at slaughter. Data were analysed using a model combining latency to and probability of outbreak at pen level. The model was analysed using bayesian statistics with the program WINBugs. Difference between treatments was evaluated based on 95% credibility intervals for relevant contrast between parameters. The incidence of pens with outbreak of tail-biting was very different between the 4 herds and was significantly less likely in pens with pigs which had ¾ of their tails removed compared to the other three treatments (Log Odds ratio compared to ¾ docking, ½: 2.8; ¼: 3.3; intact: 4.6). The odds-ratios did not differ between herds. The study also showed different growth of different segments of the tail, with largest growth at the inner segment. Thus, the only docking treatment which reduced the risk of tail-biting was the treatment where ¾ of the tail was removed, whereas the other two docking lengths were comparable with intact tails.

Which behaviour of dairy calves best indicates their level of hunger during weaning?

De Passillé, Anne Marie[1], Sweeney, Brigid[1], Weary, Daniel[2] and Rushen, Jeffrey[1], [1]Agriculture and Agri-Food Canada, Pacific Agrifood Research Centre, P.O. 1000 6947 Highway 7, Agassiz, BC, V0M 1A0, Canada, [2]University of British Columbia, Animal Welfare Program, 189 MacMillan-2357 Main Mall, Vancouver, B.C., V6T 1Z4, Canada; Annemarie.Depassille@agr.gc.ca

During weaning from milk to solid feed, dairy calves show behavioural changes consistent with hunger, but there are few empirical data relating these changes to feeding motivation. We used 32 dairy calves, housed in groups of 4 and fed milk and starter with automated feeders. We induced continuous variation between calves in total digestible energy (DE) intake during and after weaning by weaning them off milk either abruptly at 41d of age or following three gradual weaning strategies (lasting 4d, 10d or 22d) and providing free access to calf starter. We hypothesized that the purported behavioural indicators of hunger would be correlated with DE intakes. We measured time standing and lying, frequency of visits to the feeders, duration of cross-sucking and frequency of vocalizations for each calf and calculated Spearman correlations between these and DE intake (calculated from milk and starter intakes). As weaning progressed, DE intake decreased, and time spent standing, frequency of visits to the feeder, duration of cross-sucking and frequency of vocalizations increased. During weaning, DE intake was correlated with the duration of cross-sucking (r=0.56; $P<0.001$) but not with frequency of visits to the milk feeder or time spent standing. However, frequency of visits to the milk feeder tended to be negatively correlated with weight gain during weaning (r=-0.30 $P=0.08$). After weaning, DE intake was correlated with frequency of visits to the milk feeder (r=-0. 55; $P<0.001$) and time standing (r=-0.36; $P=0.02$) but not with frequency of cross-sucking or frequency of vocalization. The lack of correlation between different behavioural responses suggests that individual calves react to low energy intakes in different ways and that behaviours associated with hunger differ before and after weaning. Together these results reinforce the importance of considering multiple behavioural responses when assessing hunger in calves.

Shade seeking, feeding behaviour and social interactions in lactating camels (*Camelus Dromedarius*) on four different watering regimes

Dahlborn, Kristina, Olsson, Kerstin and Tafesse, Bekele, SLU, Anatomy, Physiology and Biochemistry, P.O. Box 7011, 750 07 Uppsala, Sweden; kristina.dahlborn@afb.slu.se

It has been believed that camels continue to eat, and thereby maintain milk production, during severe dehydration. Feed intake, milk production and behaviour were measured in seven camels in a crossover design. Repeated measurement analysis of variance (SAS software, Cary, USA) was used and data are presented as LSmeans ± SE. Water was offered once daily: (W1), every 4th (W4), 8th (W8) or 16th day (W16). The experiment was done in a paddock (24 m x 30 m) with a feeding area and a shaded area. At 7:00 hr and 17:00 hr camels were tied up, fed hay and concentrate, and hand milked. They were released between 9:00 - 12:00 and 14:00 - 17:00 and allowed to move in the paddock. At about 12:15 h the camels were weighed and then tied again and offered concentrate. At 13:00 h camels were watered according to their watering regime. The position and activity of camels were registered instantaneously every half hour between 9:00 - 12:00 and 14:00 - 17:00 on days 3, 4, 7, 8, 11, 12, 15, and 16 during each regime. The camels were offered 6 kg concentrate per day and hay was available *ad libitum*. Milk production decreased on the last dehydration day during W8, and from 2.2 ± 0.4 to 1.4 ±0.4 l during W16 ($P<0.01$). The camels spent less time feeding hay and ruminating as the dehydration time was prolonged. Concentrate intake was 5.7 ± 0.4 kg (W1), 4.8 ± 0.4 (W4), 3.6 ± 0.4 (W8), and 2.9 ± 0.4 (W16; $P<0.001$ for all comparisons). Camels spent more time in the shade the last day before watering on both W8 and W 16. Social interactions were not influenced by watering regime. In conclusion, dehydrated camels decreased food intake and milk production and spent more time in the shade.

Screening the satiating properties of dietary fibre sources in adult pigs

Souza Da Silva, Carol[1,2], Van Den Borne, Joost, J. G. C.[2], Gerrits, Walter J.J.[2], Kemp, Bas[1] and Bolhuis, J. Elizabeth[1], [1]Wageningen University, Adaptation Physiology Group, Marijkeweg 40, P.O. Box 338, 6700 AH Wageningen, Netherlands, [2]Wageningen University, Animal Nutrition Group, Marijkeweg 40, P.O. Box 338, 6700 AH Wageningen, Netherlands; carol.souza@wur.nl

Restrictedly-fed sows suffer from hunger, often signalized by increased feeding motivation. We assessed effects of three fibre sources, which may enhance satiety in different ways, on feeding motivation in gilts. Sixteen pair-housed gilts received four dietary treatments: lignocellulose (LC), pectin (PEC), resistant starch (RS), all exchanged for starch, and control (C), in four periods in a Latin square. Per period, gilts were fed a low followed by a high dose (7 days each). At 1, 3 and 7h postfeeding, feeding motivation was assessed in a runway (20g feed for walking a fixed route) and in an operant test (10g feed for wheel turns in a progressive ratio, PR1). Gilts were observed in their home-pen for 6h using 90-sec instantaneous scan sampling. Data were analyzed with mixed models including fibre type, dose and their interaction as fixed, and pen and period as random effects. In the operant test, gilts obtained more rewards 1, 3 and 7h after the PEC-diet compared to the other diets ($P<0.05$), and 1h postfeeding gilts obtained more rewards on high than low dosages ($P<0.05$). At 3h the RS-H diet tended to yield less rewards than most of the other diets, including the RS-L diet (fibre typexdose interaction, $P<0.10$). In the runway, at 1h postfeeding gilts walked faster for food when fed the PEC-diet compared with the C- and RS-diets ($P<0.05$). At 7h gilts tended to walk slower when fed the RS-diet compared to the other diets ($P<0.10$). At 3 and 7h postfeeding low-fed gilts tended to walk faster than high-fed gilts ($P=0.10$ and $P<0.10$). PEC-gilts showed more feeder-directed behaviours than RS-fed gilts ($P<0.01$), with C- and LC-fed gilts in between. In conclusion, PEC was the least satiating fibre. The RS, despite being the lowest in metabolizable energy, appeared the most satiating, possibly due to its fermentation properties.

Inter-group aggressive behaviour of free-ranging dogs (*Canis familiaris*)

Pal, Sunil Kumar, Katwa Bharati Bhaban, Life Science, Katwa Abasan, MIG (U)- 68, P.O. Khajurdihi, Dist.- Burdwan, West Bengal, 713518, Katwa, India; drskpal@rediffmail.com

The purpose of this study was to investigate the influences of sex, season and situations on the inter-group aggressive behaviour of free-ranging dogs (Canis familiaris). Observations on the inter-group aggression behaviour of 10 free-ranging dogs from two neighbouring groups were recorded in Katwa town, India. Animals were observed for 4 h /day and a total of 1,460 h over 360 days. Behavioural data were collected using ad libitum and focal animal sampling. Group home range size was maximum during the late monsoon months when the females were in oestrus. Moreover, group home range size varied with the number of dogs in a particular dog group. Overall level of aggression was higher among the females, especially among the adult females. Moreover, most of the aggressive encounters were directed towards the male dogs (t = 5.88, df = 23, $P<0.05$). Furthermore, the number of aggressive encounters varied with the number of dogs in a particular dog group. There were significant seasonal variations among the dogs in relation to the rate of aggression (F = 10.57; df = 3. 39; $P<0.05$). Male dogs were more aggressive during the late monsoon months and then during the winter months (F = 10.73; df = 3, 19; $P<0.05$), whereas female dogs were more aggressive during the winter months and then during the late monsoon months (F = 5.48; df = 3, 19; $P<0.05$). Aggressive encounters were recorded in different situations, and the situations were significantly different from each other in relation to the frequency of aggressive encounters (F = 8.96; df = 4, 49; $P<0.05$). Male dogs were more aggressive at the mating places and also at the territorial boundary, whereas female dogs were more aggressive at the feeding places and also at the nest site.

Male mink (*Mustela vison*) from enriched cages are more successful as mates

Díez-León, María and Mason, Georgia, University of Guelph, Animal and Poultry Science, 50 Stone Road East, building 70, N1G 2W1, Guelph, ON, Canada; mdiez@uoguelph.ca

Females often prefer attributes signalling health and stress-resistance in their potential mates. Mating with preferred males can also bring reproductive benefits (e.g. larger litters). We tested the hypothesis that females are therefore more motivated to mate with males with good welfare than males with poor welfare. We raised 32 male American mink from birth in either non-enriched (NEE: wire mesh breeding cage + bedded nestbox) or enriched (EE: NEE cage + two climbing towers, and an additional large cage containing, toys, a water trough, and an extra nestbox). Over two breeding seasons, females (8EE, 8NEE) were offered a choice between one EE and one NEE male (in identical cages during the test, to blind females to housing) for 4 hours/day. As predicted, females copulated with enriched males more often ($F_{1,14}=3.20$, $P<0.05$) and for longer ($F_{1,14}=4.10$, $P<0.05$). Phenotypically, NEE and EE males differed as follows. EE males were heavier ($F_{1,25}=7.93$; $P<0.01$) and tended to be more symmetrical ($F_{1,29}=3.27$, $P=0.08$). They had relatively bigger thymuses ($F_{1,23}=4.64$; $P<0.05$) and spleens ($F_{1,23}=8.36$, $P<0.01$), suggesting better immune status; and lower fecal glucocorticoid concentrations ($F_{1,29}=8.33$, $P<0.01$). EE males also had relatively larger brains ($F_{1,22}=5.57$; $P<0.05$), and performed less stereotypic behaviour ($F_{1,30}=26.97$, $P<0.001$). From all these traits, lower levels of stereotypic behaviours predicted more ($F_{1,15}=5.46$, $P<0.05$) and longer ($F_{1,15}=7.71$, $P<0.05$) copulations; and large male body weights had similar effects ($F_{1,15}=3.46$, $P=0.08$; $F_{1,15}=4.59$, $P<0.05$). These two traits were inter-related, body weight and stereotypic behaviour negatively correlating ($F_{1,28}=5.48$, $P<0.05$). Therefore, the physical and/or psychological correlates of stereotypic behaviour seemed to underlie the lower mating success of the NEE males. Overall, these results show that high welfare males with little stereotypic behaviour are more successful, being more attractive, more sexually motivated, or both. More generally, they also potentially help explain why captive wild animals sometimes have poor reproductive success.

Management measures to reduce sexual behaviour in male pigs

Vermeer, Herman, Houwers, Wim and Van Der Peet-Schwering, Carola, Wageningen UR, Livestock Research, P.O. Box 65, 8200 AB Lelystad, Netherlands; herman.vermeer@wur.nl

The Dutch pig sector moves towards non castration in male piglets. This does not only require a solution to reduce and identify boar taint but also one to reduce typical 'male behaviour' like mounting and aggression causing a threat to both animal welfare and performance. The objective was to reduce sexual behaviour and aggression in male pigs by management measures. Two experiments were conducted in enriched pens of 29 m^2. In both experiments skin injuries (score 0-5), lameness (score 0-2) and mounting behaviour (mounting (attempts) per pen per snapshot) were measured and analysed using REML in Genstat. In the first experiment we used 12 pens with 18 pigs each to test the difference between boars and barrows, straw and rubber as lying area and 3 and 6 eating places. Boars showed more skin lesions (0.64 vs. 0.48; $P=0.059$) and mounting behaviour (0.29 vs. 0.12; $P<0.007$) than barrows, but no difference in lameness (0.41 vs. 0.30; $P=0.27$). Straw and an additional feeder did not reduce mounting behaviour and an additional feeder did not reduce lameness (0.40-0.31; $P=0.47$) and tended to reduce skin lesions (1.88-1.48; $P=0.098$). In the second experiment we used 8 pens with 15 boars each for a 2 × 2 factorial design to test the effect of additional sugar beet pulp pellets (SBP) and the presence of a dummy to mount. SBP did not reduce lesions, lameness and mounting behaviour. The dummy resulted in less lameness (0.23 vs. 0.32; $P=0.018$) and did not reduce skin lesions (0.73 vs. 0.81; $P=0.597$), however it didn't effect mounting behaviour. It can be concluded that mounting behaviour is not reduced by the tested measures. The dummy reduced lameness but served more as a barrier than as a mounting device. A substantial reduction of male pig behaviour must probably be found in a combination of measures.

Important considerations for the use of older adult dogs in ethological research

Salvin, Hannah[1], McGreevy, Paul[1], Sachdev, Perminder[2,3,4] and Valenzuela, Michael[2,4], [1]University of Sydney, Faculty of Veterinary Science, (B19) Camperdown, NSW 2006, Australia, [2]University of NSW, School of Psychiatry, Randwick, NSW 2031, Australia, [3]Prince of Wales Hospital, Neuropsychiatric Institute, Randwick, NSW 2031, Australia, [4]University of NSW, Faculty of Medicine, Brain and ageing research program, Randwick 2031, Australia; hannah.salvin@sydney.edu.au

The dog is rapidly gaining importance as an ethological subject. Researchers have traditionally clustered physically mature dogs as 'adults', ignoring potentially confounding age-related behavioural change that may arise as early as 6yrs of age. Our study aimed to isolate behavioural responses affected by age to allow ethologists to control for these confounders. We surveyed owners of dogs ≥8 years of age (n=1,100) from 11 countries. Questions (n=83) related to frequency of a response or level of deterioration over a 6-month period. Dogs with behavioural profiles indicative of cognitive dysfunction were excluded. Eligible responses (n=826) were arranged according to age groups: <10yrs; 10-12yrs and; >12yrs. Regression identified significant age effects from 18 responses (difference across age: 4.5%-30.3%, $P<0.001$-0.038). Of particular interest; was a 21.3% decrease in dogs showing reliable response to commands with age ($P<0.001$) and a 30.3% increase in dogs having difficulty finding dropped food ($P<0.001$). Significant age effects on deterioration over the preceding six months was evident in 21 responses (difference across age: 3.5%-25.7%, $P<0.001$-0.033).With increasing age, there was an 8.7% increase in dogs showing an increased latency to learn ($P=0.028$). There was a 12.1% increase in dogs taking longer to eat meals ($P=0.001$) and a 19.6% increase in dogs spending less time active ($P<0.001$). Numerous responses pertinent to canine ethological research were shown to be age-sensitive. Responses to do with learning and food motivation may influence research reliant on training or food-based rewards. Field-based research requiring significant levels of activity may also be impacted. It is critical that canine ethologists embrace the effect of age, especially in groups of companion dogs that are from various breeds. An understanding of age-sensitive responses should underpin ethological studies involving 'adult' dogs.

Validation of accelerometers to automatically record sow postures and stepping behaviour

Ringgenberg, Nadine[1,2], Bergeron, Renée[3] and Devillers, Nicolas[1], [1]Agriculture and Agri-Food Canada, Dairy and Swine R&D Centre, 2000 College Street, P.O. Box 90, STN Lennoxville, J1M 1Z3 Sherbrooke, QC, Canada, [2]University of Guelph, Animal and Poultry Science, 50 Stone Road East, N1G 2W1 Guelph ON, Canada, [3]University of Guelph, Animal and Poultry Science - Alfred Campus, 31, rue St-Paul - C.P. 580, K0B 1A0 Alfred ON, Canada; Nadine.Ringgenberg@agr.gc.ca

Our aim was to develop an automated method for detecting postures and stepping behaviour in sows using accelerometers (Pendant G Acceleration Data Loggers, Onset Computer Corporation). In a first study, the simultaneous application of two accelerometers was tested on 12 multiparous sows to detect standing, sitting, lying ventrally and lying laterally. The sows were housed in gestating stalls (n=4) or pens (n=4), farrowing crates (n=2) or pens with straw (n=2). One accelerometer was fastened to a hind leg and the other on the back of the sow. They recorded acceleration on 3 axes every 5 sec for 6 h. Data from the two accelerometers were converted into degrees of tilt and used in conjunction to discriminate between postures; they were then compared to video recordings. Based on video observations, sows spent an average time of 23±0.1% standing, 25±0.2% lying ventrally, 48±0.3% lying laterally and 4±0.1% sitting. Sensitivities (the extent to which the accelerometers correctly detect each posture) for standing, lying ventrally, lying laterally and sitting were 100±0.01%, 94±0.04%, 91±0.2% and 50±0.4%, respectively. Specificity (the extent to which the accelerometers correctly identify true negatives) was above 90% for all postures. The second study tested accelerometers for the quantification of stepping behaviour (a potential predictor of lameness) at feeding on 10 sows. The sows were housed in gestating stalls (n=7) or pens (n=3) and had one accelerometer fastened to a rear leg. The data loggers recorded acceleration on the vertical axis 10 times per sec for 30 min starting at the time of feeding. The accelerometer data was compared to video observations, 1,449 steps were assessed in total. The average sensitivity was 94±0.04% with an error of 5±0.03%. In conclusion, accelerometers can be successfully used to estimate postures and stepping behaviour in sows.

Variation in growth, feed intake and behaviour of sole

Mas Muñoz, J[1,2], Schneider, O[2], Komen, H[1] and Schrama, J[1], [1]Wageningen University, P.O. Box 338, 6700AH Wageningen, Netherlands, [2]IMARES, P.O. Box 77, 4400AB Yerseke, Netherlands; julia.mas@wur.nl

The goal of this study was to assess the relationship between feed intake, growth and (non-) feeding behaviour of 16 individually housed sole (Solea solea) during a 4-week period. Fish were hand fed three times a day (8:00, 12:00 and 17:00 h) for 20 minutes until satiation. For each individual, behaviour was recorded during 3 minutes by direct observations twice a day in between meals at Days 8, 10, 13, 15, 16 and 24. Total swimming activity (% of observation), frequency of burying (bouts of quick wave movements) and frequency of escaping (bouts of swimming vertically at the surface with head now and then out of the water) were recorded. At the beginning and at the end of the growth period, two sequential behavioural tests (15 min each) were performed. During the 'Novel Environment Test' reaction of fish to a new environment was analyzed whereas during the 'Light avoidance test' reaction of fish to a sudden increase in light intensity was analyzed. Individually housed fish still exhibited pronounced variation in feed intake (23%), growth (25%), feeding consistency (27-55%), swimming activity (97%), and behaviour during tests (141-170%). Variation in feed intake between days and between meals within days (%) were negatively correlated with total feed intake (r=-0.65, $P<0.01$ and r=-0.77, $P<0.001$). Swimming activity was positively correlated with feed intake (r=0.66, $P<0.01$) and growth rate (r=0.58, $P<0.05$). Moreover, proactive sole: (1) fish which reacted with a lower latency time to move after its introduction in a novel environment had higher feed intake (r=-0.55, $P<0.05$) and growth (r=-0.66, $P<0.01$); (2) fish which escaped during the light avoidance test tended to show higher feed intake ($P<0.1$) and had higher growth rate ($P<0.05$). In conclusion, feeding consistency, swimming activity in the tank, and behavioural responses to novel tests are related to feed intake and growth of sole.

Behaviour in two group sizes of beef cattle kept in a mobile outdoor system during winter

Wahlund, Lotten[1], Lindgren, Kristina[1], Salomon, Eva[1] and Lidfors, Lena[2], [1]JTI - Swedish Institute of Agricultural and Environmental Engineering, P.O. Box 7033, SE-750 07 Uppsala, Sweden, [2]Swedish University of Agricultural Sciences, Department of Animal Environment and Health, P.O. Box 234, SE-532 23 Skara, Sweden; lotten.wahlund@jti.se

The aim of this study was to evaluate behaviour in two different group sizes of pregnant beef cattle in relation to their use of resources, and how much and for which types of behaviour a new type of weather shelter was used. The study was carried out during wintertime on a commercial farm in Western Sweden. A large group (LG, n=34) and a small group (SG, n=17) were kept outdoors with access to expandable areas of arable grassland (LG 1,500-4,500m^2, SG 750-2,250m^2). SG had access to one and LG to two tents, where each tent (78m^2) had a full canvas roof, wind netting on three sides and deep straw bedding. Number of feeding places and tent-area per heifer were equal in both groups. Feed-area and feed-racks were moved weekly in both groups while the water-area was stationary. Behavioural observations were performed by three observers simultaneously in both groups during 2 consecutive days/week (Day1: 11-12am, 13-17pm, Day2: 7-11am, 12-13pm) repeated three weeks. The number of heifers in each defined sub-area and type of behaviour were recorded every 5 minutes. The heifers were recorded to be most often in the feed-area (LG: 71.5%, SG: 76.1%) and thereafter in the tent-area (LG: 25%, SG 20.4%). The heifers were recorded to be in the water-area somewhat more often in LG (2.1%) than in SG (1.4%). The most common observed types of behaviour were eating (LG: 54.3%, SG: 53.2%), standing (LG: 19.7%, SG: 19.5%) and lying (LG: 18.5%, SG: 15.9%). Lying in the tent was performed equally often in the two group sizes (15.7%), whereas heifers in LG appeared to stand more in the tent (7.4%) than heifers in SG (3.3%). In this case-study the heifers in the large group appeared to have similar access to weather shelter and feed as in the small group.

Group housing of horses under Nordic conditions: strategies to improve horse welfare and human safety

Keeling, LJ[1], Søndergaard, E[2], Hyyppä, S[3], Bøe, K[4], Jørgensen, G[4], Mejdell, C[5], Ladewig, J[6], Särkijärvi, S[3], Jansson, H[3], Rundgren, M[1] and Hartmann, E[1], [1]Swedish University of Agricultural Sciences, Animal Environment and Health, P.O. Box 7068, 750 07 Uppsala, Sweden, [2]AgroTech, Animal science, Aarhus, 8200, Denmark, [3]Agrifood Research, Ypäjä, 32100, Finland, [4]Norwegian University of Life Sciences, Department of Animal and Aquacultural Sciences, P.O. Box 5003, 1432 Ås, Norway, [5]National Veterinary Institute, Oslo, 0454, Norway, [6]University of Copenhagen, Faculty of Life Sciences, Frederiksberg C, 1870, Denmark; linda.keeling@hmh.slu.se

Despite many welfare advantages, owners are often concerned about keeping horses in groups. The main concerns seem to be risk of injuries when mixing horses, the difficulty of separating a horse from the group, fear reactions and the consequently potentially dangerous situation for humans moving within a group of horses. These concerns were addressed by observations in the home paddock of groups of horses matched by group size (range 3-8) and breed (Icelandic, Warmblood, other) to allow comparisons of three different group compositions: 'single age' versus 'mixed age' (28 groups), 'mixed sex' versus 'single sex' (24 groups) and 'stable' versus 'unstable' group membership (15 groups). Records of injuries were taken directly after mixing and 4-6 weeks later. Other tests were carried out at least 2 weeks later. There was no difference between treatments in the reactivity of the horses to a standardized reactivity test (suddenly moving novel object), or in the number or severity of injuries when mixing horses. However, it was easier to remove a horse (scored according to ease of approach, haltering and following out from the paddock) from a mixed sex group than if the group was all female ($P<0.001$) or all male ($P<0.01$). Irrespective of treatment, there were significant breed differences in that groups of Icelandic horses had fewer and less severe injuries than groups of other breeds and the injury was more likely to be on the rump whereas in groups with other breeds it was more likely to be at the head ($P<0.01$). Icelandic horses also reacted least in the reactivity test ($P<0.001$), but groups with other breeds calmed down more quickly after exposure to the moving novel object ($P<0.05$). In summary, there was little or no support for the original areas of concern against keeping horses in groups.

Effects of previous housing in large groups on behavior of growing pigs at mixing

Wang, Lihua[1,2] and Li, Yuzhi[2], [1]Qingdao Agricultural University, College of Animal Science and Technology, Qingdao Agricultural University, College of Animal Science and Technology, Qingdao, Shandong Province, 266109, China, [2]University of Minnesota, West Central Research and Outreach Center, WCROC, 46352 State Highway 329, Morris, MN, 56267, USA; yuzhili@morris.umn.edu

A study was conducted to investigate effects of previous housing in large groups on behavior of growing pigs at mixing. A total of 216 pigs, 108 derived from large groups and 108 from small groups were used. Pigs from large groups mingled in groups of 80 pigs at 10-day old in a group-farrowing barn. Pigs from small groups originated from a conventional nursery barn with 9 pigs per pen. At 8-wk old, pigs (23 ±3.4 kg) were allocated to 24 pens of 9 pigs (5 males and 4 females), 12 pens per treatment. The familiarity among pigs within a pen was the same between the two groups. Behavior was video-recorded for 24 hr after mixing in 6 pens per treatment. The image was viewed continuously during the initial 2 hr to record duration and frequency of fighting among the pigs. Time spent on lying, standing, eating, and drinking was estimated by scan sampling at 5 min intervals for 24 hr. The body weight and feed intake of pigs in 24 pens was monitored every 2-wk for 22 wk. Data were analyzed by using the Glimmix Procedure of SAS. Compared with pigs from small groups, pigs derived from large groups were less aggressive (duration of fighting=281 vs. 1941 sec/2hr/pig, SE=174.7; $P<0.001$; number of fights=14.7 vs. 40.7 fights/2hr/pig, SE=7.64; $P<0.05$), spent more time lying (82.9% vs. 78.6%, SE=0.95; $P<0.01$), and less time standing (7.1% vs. 9.0%, SE=0.59; $P<0.05$) and eating (8.7 vs. 10.9%, SE=0.52; $P<0.01$) at mixing. The growth performance data showed that pigs derived from large groups were heavier (106.9 vs.103.0 kg, SE=0.59; $P<0.01$) than pigs from small groups at wk 22. These results indicate that pigs previously housed in large groups were tolerant to unfamiliar pigs, which could alleviate mixing-induced stress and improve growth performance.

Changes on the response to social separation in one year old bucks
Ruíz, Osvaldo, González, Francisco, Méndoza, Niza, Medrano, Alfredo, Paredes, Marisol, Terrázas, Angélica and Soto, Rosalba, Facultad de Estudios Superiores Cuautitlán, Universidad Nacional Autónoma de México, Secretaria de Posgrado e Investigación, Carr. Cuautitlán-Teoloyucan Km 2.2 Cuautitlán Izcalli, 54714, Mexico; rosaneopri@yahoo.com.mx

Domestic goats are gregarious animals. In range conditions, adult males and females maintain separate social groups, except for the seasonal activities around mating. In different management systems, the bucks are isolated from the rest of the flock in individual pens. However, little is known if social separation produces some degree of disturbance or distress that could alter the welfare of these animals. This work was carried out to study changes in social behaviour in one standarized test in 20 juvenile bucks which were one year old. Each subject was placed into a holding pen (2 × 2m) within the main flock farm yard. In the first part, the male stayed 5 min surrounded by its conspecifics; in the second part the male stayed alone for 5 min. Vocalizations, sniffing objects, locomotion, eliminations and jumps were recorded. An agitation index was made considering all these variables, and response to social separation was constituted by differences in each part of the test. The bucks performed a lower number of high bleachs (1.7±1.1 vs. 85.7± 7.6), low bleachs (0.6± 0.3 vs. 8.0 ± 3.2) and sniffing on objects (0.5± 2.2 vs. 5 ±1.1) when kept with conspecifics compared to when kept alone ($P<0.004$, Wilcoxon test, in all cases). However, the locomotion did not differ ($P<0.06$, Wilcoxon test). Finally, the agitation index was different in the first part with conspecifics in contrast to the second part without the main flock (25.0± 2.3 vs. 118.3± 9.1, $P<0.001$, Wilcoxon test). In conclusion, young Bucks display an intensive response to the social separation test. Project supported by UNAM-FES-Cuautitlán, CONS 201 and PAPIIT IN207508-2.

Heart rate and urinary hormone concentration in dogs: diurnal variation and relation to reactivity

Axel-Nilsson, Malin[1] and Hydbring-Sandberg, Eva[2], [1]Swedish University of Agricultural Sciences, Department of Animal Environment and Health, P.O. Box 7068, SE-750 07 Uppsala, Sweden, [2]Swedish University of Agricultural Sciences, Department of Anatomy, Physiology and Biochemistry, P.O. Box 7011, SE-750 07 Uppsala, Sweden; malin.axel-nilsson@hmh.slu.se

This study aimed to investigate relations between canine basal levels of heart rate, adrenaline, noradrenaline, cortisol and testosterone in the home environment, and reactivity during a fear test (in the lab). 41 privately owned dogs of different breeds (n=20), gender ($\male$=26, $\female$=15), and ages (7 months-12years) were used. A sub sample of ten male dogs was exposed to an unfamiliar dog and two artificial dogs in a fear test. Hormonal/creatinine ratios were analyzed in naturally voided urine samples collected in the dogs' home environment morning and evening, and at the test site immediately before and 30 minutes after the test. Heart rate was registered by a telemetric device (Polar) earlier used in dogs. ANOVA, Mixed Model, SAS, was used for statistical analysis. Average heart rate of all dogs during a 24-hour registration was 80 ± 2 beats per minute (bpm) and did not differ between gender, but male dogs reached higher maximum heart rate (206 ± 4 bpm) than bitches (188 ± 6 bpm) ($P\leq0.001$). The ratios of adrenaline and testosterone were significantly higher in the evening compared to morning samples in all dogs ($P\leq0.05$). Males had higher testosterone ratio than bitches ($P\leq0,001$), morning and evening. Direct observations and assessment of video recordings were used for behavioural analyses of the fear test. The ten dogs were classified as more or less reactive (aggression and/or fear). The adrenaline ratio was increased after the test (n=10, $P\leq0,05$). Analyses further revealed that the five more reactive dogs had higher heart rate and testosterone ratios than the other five (139 ± 8.8 bpm vs. 110 ± 5.7 bpm; testosterone= 82.1 ± 20.1 vs. testosterone= $34.9\pm7.8\times10^{-6}$; $P\leq0.05$) during/after test. In conclusion, basal levels of heart rate and concentrations of adrenaline and testosterone were affected by several factors including reactivity. Further studies are required to find out whether there are physiological differences between ages and breeds in dogs.

Effect of manner of approach by an unfamiliar person on behaviour and eye/ear temperatures of dogs (*Canis lupus familiaris*)

Fukuzawa, Megumi, Kanaoka, Kanako, Ema, Tsukasa and Kai, Osamu, Nihon University, College of Bioresource Sciences, Department of Animal Science and Resources, 1866 Kameino, Fujisawa, Kanagawa, 252-0880, Japan; fukuzawa.megumi@nihon-u.ac.jp

The dogs can adapt their responses to human social cues; however, there have been few evaluations of the physiological reactions of dogs to such cues. In this study, five pet dogs were used as subjects to investigate the both behavioural and physiological response to approach manner of an unfamiliar person. We observed two types of approach (friendly and expressionless approach) by people of either sex, and the tests were filmed using a video camera. The dogs' body and eye temperatures were measured with a thermal video system, and their ear temperatures were measured with a Vet-Temp electronic ear thermometer. The total durations of the two behavioural categories [social behaviour toward approaching person] (average; 42.13 sec.) and [social behaviour toward experimenter] (42.88 sec.) with a friendly approach were significantly longer than with an expressionless approach (14.75 sec: 2.0 sec.) (ANOVA, $P<0.001$; respectively). With the friendly approach, the total duration of [social behaviour toward approaching person] was longer than [refusal behaviour toward approaching person] (15.0 sec.) (Tukey, $P<0.05$). Each type of behaviour was not affected by the sex of the approaching person. The thermal images for eye and ear temperatures were positive correlated with each other (Pearson's correlation: $r = 0.65$, $P<0.01$ for the time point at which the approach began; $r = 0.66$, $P<0.01$ for the time point at which the approach ended). There were no significant differences in the thermal images or ear temperatures between the two measurement points. These results suggested that the dogs quickly recognised the approach person's attitude. They might be also suggested that the dog's behavioural response was influenced by the approach manner rather than by the sex of the approaching person. There were no significant short-term changes in body temperature over a short time. Even if the measurement of ear temperature is difficult, it is possible to estimate eye temperature by using a thermal video system.

Peckers like to peck: selection for feather pecking leads to increased pecking motivation in laying hens

De Haas, Elske[1], Nielsen, Birte[2], Buitenhuis, Bart[3] and Rodenburg, Bas[4], [1]Wageningen University and Research Centre, Adaptation Physiology Group, Marijkeweg 40, 6709PG, Wageningen, Netherlands, [2]Aarhus University, Department of Animal Health and Bioscience, Blichers Allé, DK8830, Tjele, Netherlands, [3]Aarhus University, Department of Genetics and Biotechnology, Blichers Allé, DK8830, Tjele, Denmark, [4]Wageningen University and Research Centre, Animal Breeding and Genomics Centre, Marijkeweg 40, 6709PG, Wageningen, Netherlands; elske.dehaas@wur.nl

Feather pecking (FP) – although thoroughly studied – is still a major welfare problem in laying hens. It is known that FP is related to foraging behaviour, but it remains unclear whether FP substitutes for foraging or whether feather peckers continue to show high levels of foraging. To assess the relationship between FP, foraging and food preference, we studied 16 low FP hens and 16 high FP hens originating from divergent selection lines on FP behaviour. Hens were tested individually in a foraging plus-maze, in which one of four food-items was displayed in each arm: (1) regular food-pellets, (2) grass, (3) downy back feathers, and 4) dead mealworms in wood-shavings. Hens were first habituated to the empty maze and hereafter trained thrice in the maze with food-items inside (10 minutes/trial). The fourth trial was set as test trial, in which we recorded eating time, eating bouts and amount eaten of each food-item. Behaviour was recorded continuously with a PSION recording device. Data were analysed using ANOVA, testing effects of line. Both lines spent most time eating worms compared to eating grass and feathers (GLM contrast: $F_{1,29}$= 9.14, $P<0.01$; $F_{1,29}$= 10.55, $P<0.01$). However, in comparison to low FP hens, high FP hens ate worms faster (1.02 w/min ±0.3 vs. 0.24 w/min ± 0.18; $F_{1,31}$ = 3.99, $P<0.05$) and tended to have more worm eating bouts (3.38± 0.02 vs. 0.87± 0.00; $F_{1,31}$= 3.87, P= 0.06). When first subjected to the maze, high FP hens also displayed more exploratory pecks (8± 3 vs. 1± 0.25; $F_{1,28}$= 6.53, P= 0.02) than low FP hens. Laying hens are attracted to mealworms and like to work for food, which can explain their preference for mealworms hidden in wood-shavings. Selection for FP led to increased foraging pecking in the high FP hens, shown by their high eating rate, eating bouts and exploratory pecking, suggesting that FP co-exists with rather than substitutes for foraging.

Social space in goats: individual distance during resting and feeding in homogeneous and heterogeneous groups of young and adult goats

Ehrlenbruch, Rebecca, Andersen, Inger Lise, Jørgensen, Grete Helen Meisfjord and Bøe, Knut Egil, Norwegian University of Life Sciences, Department of Animal and Aquacultural Sciences, P.O. Box 5003, 1432 Ås, Norway; rebecca.ehrlenbruch@umb.no

The aim of the experiment was to investigate the individual distance during resting and feeding in homogenous vs. heterogeneous groups of young and adult goats. Thirty-six young (<2 years) goats and thirty-six adult (≥2 years) goats of the Norwegian dairy breed were allotted into 6 groups with young goats (4 goats), 6 groups with adult goats (4 goats) and 6 heterogeneous groups (2 young goats and 2 adult goats). Each group was kept in an experimental pen (12.0m × 2.0m) for 3 days. A solid resting platform (12.0m × 0.6m) and a 12.0m long feed barrier was running on each long side, so that the goats could maximize the distance between each other. After two days of getting accustomed to the experimental pen, a 24 hour video recording was conducted. The distance between all four goats in the group (6 possible pairs) when resting was scored every 10 minutes for 12 hours (from 06:00 p.m. to 06:00 a.m., quiet period of the 24 h) and distance between goats when feeding was scored every second minute for 2 hours after each feeding (09:00 a.m. and 05:00 p.m.). The individual distance was significantly lower for the heterogeneous groups during resting (mean ± SE; 1.8 ± 0.2 m) than the homogenous groups (young: 2.4 ± 0.3 m; adult: 2.7 ± 0.3 m) ($P<0.01$). During feeding there was no such difference (young: 1.7 ± 0.1 m; heterogeneous: 2.9 ± 0.2 m; adult: 2.7 ± 0.3 m). The heterogeneous groups had significantly larger individual distance when feeding than when resting ($P<0.001$), while type of activity did not affect individual distance in homogenous groups of young and adult goats. In conclusion, individual distance was influenced by age composition and type of activity.

Behaviour, heart rate and cortisol in goat kids reared with or without mother

Winblad Von Walter, Louise, Hydbring-Sandberg, Eva and Dahlborn, Kristina, Swedish University of Agricultural Sciences, Anatomy, physiology and biochemistry, P.O. Box 7011, 75007 Uppsala, Sweden; louise.winblad@afb.slu.se

The purpose of this study was to investigate how different rearing methods affect heart rate, saliva cortisol concentrations and behaviour in goat kids. After the colostrum period, ten kids were separated from their mothers and reared together (Group 1). Six kids were separated from their mothers at daytime between morning and afternoon milkings (Group 2) and six kids were held together with their mothers (Group 3). At 2 weeks and 2 months of age, behavioural observations were made in the different groups and an isolation test with subsequent reunion was performed. The kids that were permanently separated from their mothers spent more time lying down close to another kid than the other groups at 2 weeks of age. At 2 months of age, kids in group 1 showed more active behaviours than kids in group 2. There were no differences in heart rate between treatments at the 2 weeks isolation test. The reunion showed a higher heart rate for group 1 (210±1) than group 2 (206±1) and group 3 (204±1), $P \leq 0.05$. At 2 months isolation test, the heart rate was higher for group 1 (184±1) than group 3 (178±1) and group 2 (176±1), $P \leq 0.05$. The reunion showed significant differences between all treatments with the highest for group 1 (178±1), followed by group 3 (171±1) and group 2 (164±1), $P \leq 0.05$. There were no significant differences between the groups in saliva cortisol concentrations. At 2 weeks isolation test, the vocalization frequency was highest for group 2, followed by group 3 and group 1. At 2 months there were no differences between treatments. In conclusion, young kids reared with their mothers tried to communicate during isolation and thereby they reduced their stress level as indicated by the lower heart rate.

Heart rate increases in puppies during behaviour tests and differs between litters

Hydbring-Sandberg, Eva, Ericsson, Maria, Enfeldt, Maria and Svartberg, Kenth, Faculty of Veterinary Medicine and Animal Science, Swedish University of Agricultural Sciences, Department of Anatomy, Physiology and Biochemistry, P.O. Box 7011, SE - 750 07 Uppsala, Sweden; eva.sandberg@afb.slu.se

Many dog breeders use different tests in order to select the right puppy to the right owner. The purpose of this study was to register heart rate in seven week old puppies to find out if heart rate varies during a test and if there are variations in heart rate between litters. Thirty-six privately owned puppies from five different litters, i.e. two Labrador Retrievers, one Golden Retriever, one Groenendael and one Border collie, participated in the study. Each test was performed at two consecutive days in a mobile test arena that was built up in the breeders' home. On the first day the puppies were tested for restraint, isolation and metallic noises. On the second day they were exposed to a novel moving object and hidden food. The puppies were videotaped and heart rate was measured by phonendoscope with the puppy gently restrained during 15 s before and after each test. Heart rate increased significantly ($P \leq 0.05$) after isolation (from 161 $\pm$ 3 to 173 $\pm$ 3 beats/min), a metallic noise (from 160 $\pm$ 3 to 174 $\pm$ 3 beats/min), a novel moving object (from 168 $\pm$ 3 to 180 $\pm$ 3 beats/min) and hidden food (from 169 $\pm$ 3 to 179 $\pm$ 2 beats/min) but not during restraint (168 $\pm$ 3 to 174 $\pm$ 3 beats/min). The first control registrations did not differ during the two different days (168 $\pm$ 3 vs. 168 $\pm$ 3 beats/min). Alterations in heart rate were similar during the different tests in all litters but basal levels varied significantly ($P \leq 0.05$). Means $\pm$ SEM, ANOVA, Mixed Proc., SAS. In conclusion, all tests except restraint increased heart rate in the puppies which show that the puppies were affected by the tests. In addition, there were heart rate differences between litters, which is worth attention in future studies.

Morphological intestinal health and its implications to tail biting behavior

Palander, Pälvi[1], Brunberg, Emma[2], Munsterhjelm, Camilla[2], Sihvo, Hanna-Kaisa[3], Valros, Anna[1] and Heinonen, Mari[1], [1]University of Helsinki, Department of production animal medicine, P.O. Box 57, 00014 University of Helsinki, Finland, [2]Swedish university of agricultural sciences, Department of animal environment and health, P.O. Box 7068, 75007 Uppsala, Sweden, [3]University of Helsinki, Department of basic veterinary sciences, P.O. Box 66, 00014 University of Helsinki, Finland; palvi.palander@helsinki.fi

Tail biters have been suggested to be smaller than other pigs, possibly due to nutritional deficiencies or early life damage leading to poor utilization of feed. To study the connections between tail biting behavior and gut health, we examined the morphology of the gut mucosal cell wall. The gut wall of healthy pigs has higher villi with less and lower crypts than the gut wall of sick, malnourished or recently weaned pigs. Age- and gender-matched quartets of fattening pigs (n=55) were chosen according to their tail biting behavior using direct observations on a Finnish tail biting problem farm. Experimental groups were tail biters (TB), victims (V), control pigs in the same (C_{tb}) or in another pen without tail biting (C_{no}). After euthanisation formalin-preserved, paraffin-embedded and stained 4 μm jejunal samples were taken 50 and 100 cm past the bile duct. Villus hight, crypt depth and villus to crypt ratio were analyzed morphometrically. Data was tested against behavioural group using ANOVA and Bonferroni´s test for multiple comparisons. Age was included in the model. Values were adjusted to cube root of body weight and followed loglinear distribution. Villi were shorter ($P=0.01$) in C_{tb} compared to TB (398.4 ± 24.61 vs. 530.0 ± 22.43) and C_{no} (538.8 ± 32.85). The fact that C_{no} pigs had higher villi than C_{tb} might be a consequence of a less stressful pen environment and less competition for food. This can even be true for C_{tb} compared to TB because avoiding biters in feeding situations might cause decreased feed intake and villi shortening. There seems to be a connection between gut wall structure and tail biting behaviour but the causal mechanism of this is unknown.

Effect of straw on the behavioural, physiology and productive variables in feed lot lambs

Lemos Teixeira, Dayane[1], Miranda-De La Lama, Genaro[1], Villarroel, Morris[2], Garcia, Silvia[1], Escós, Juan[1] and María, Gustavo[1], [1]University of Zaragoza, Faculty of Veterinary/Departament of Animal Production, C/ Miguel de Servet, 177, 50013 Zaragoza, Spain, [2]Polytechnic University of Madrid, Ciudad Universitaria s/n, 28040 Madrid, Spain; teixeira@unizar.es

Lamb production systems in Spain are changing to new scheme including an intermediate step between the farm and the abattoir at corporative classification centres. Since the cost of straw has increased, lamb feed lot managers aim to avoid the use of straw during this stage. The study analyzed the use of straw as an enrichment material in lambs on the behavioural, physiology and productive variables. We used 24 male lambs (17.2 ± 0.2 kg and 60 d old). The animals were either housed in pens ($5.6m^2$) with commercial concentrate and water with straw (either to lie or to eat) or without straw at all. Two replicates were performed per group with six lambs each (2 housing type x 2 replicates x 6 lambs). On days 1, 7, 14, 21 and 28 the lambs were observed using cameras (08:00 to 20:00 h). One day after each observations, blood samples were taken and lambs were weighted. The behaviours are presented as mean events per animal per day. All dates were analysed using a mixed model with repeated measures including the effect of type of housing and the random effect of lamb. Lambs without straw showed more stereotypes (43.9 ± 1.6 vs. 26.5 ± 1.6; $P<0.01$), more aggressive interactions (29.6 ± 2.1 vs. 22.1 ± 2.1; $P<0.01$) and more affiliative interactions (39.3 ± 2.5 vs. 28.6 ± 2.5; $P<0.01$) than lambs with straw. Lambs without straw presented higher ($P<0.05$) values of cortisol (5.90 ± 1.09 vs. 2.15 ± 1.09) and lactate (24.95 ± 1.19 vs. 21.39 ± 1.19) than lambs with straw. However, lambs without straw showed lower ($P<0.05$) values of the white blood cells (6.09 ± 0.24 vs. 7.13 ± 0.24), red blood cells (9.72 ± 0.11 vs. 10.20 ± 0.11) and haemoglobin (10.88 ± 0.12 vs. 11.28 ± 0.12) than lambs with straw. Live weight, daily gain and conversion index were not affected. The results suggest that straw provision can be considered as an important tool of the environmental enrichment to improve animal welfare.

Are welfare of dairy cows and young stock related?

De Vries, Marion[1], Bokkers, Eddie A.M.[1], Dijkstra, Thomas[2], Van Schaik, Gerdien[2] and De Boer, Imke J.M.[1], [1]Wageningen University, Animal Production Systems, P.O. Box 338, 6700 AH Wageningen, Netherlands, [2]Animal Health Service, P.O. Box 9, 7400 AA Deventer, Netherlands; marion.devries@wur.nl

Raising young stock and dairy cows on the same farm may imply that the welfare of these groups are associated, e.g. because of similarities in management. The aim of this research was to identify associations between some welfare indicators of cows and young stock in dairy farms. Behavioural and clinical health data of loose housed dairy cows (n=1,013) and female young stock (n=1233) were collected in a 3-months period on 23 randomly selected dairy farms in the Netherlands. Farms housed between 35 and 177 dairy cows and between 18 and 122 young stock. Young stock was assessed only when housed in groups and divided in four age categories: (1) 0-3, (2) 3-6, (3) 6-12, and (4) 12-18 months. Behavioural response to humans was tested by measuring avoidance distance at the feeding rack (ADF). Clinical parameters included individual animal recordings on body condition, locomotion, cleanliness, respiration, diarrhea and integument alterations. Associations of dairy cows with each of the four age categories in young stock were analyzed on farm level with Spearman rank correlation tests. ADF in dairy cows was not significantly associated with ADF for any of the age categories in young stock. Also, no associations between dairy cows and any of the age categories in young stock were found for locomotion, cleanliness, respiration, and diarrhea. The percentage of lean cows (5.0±6.4) was associated with the percentage of lean young stock in age category 1 (8.3±15.8; r=0.60, $P<0.01$), and showed trends with young stock in age categories 3 (7.5±11.9) and 4 (5.9±13.2) ($P<0.1$). Number of integument alterations per cow per farm (3.1±2.3) was associated with the percentage of young stock with lesions in age categories 2 (4.7±13.0) and 3 (6.0±10.5) (r=0.49 and 0.55, $P<0.05$). Welfare parameters in this study showed few associations between dairy cows and young stock.

Effects of pre- and postweaning housing conditions and personality on feeding-related behaviour and growth of newly weaned piglets

Ooms, Monique, Oostindjer, Marije, Van Den Brand, Henry, Kemp, Bas and Bolhuis, J. Elizabeth,
Wageningen University, Adaptation Physiology Group, Marijkeweg 40, 6709 PG Wageningen,
Netherlands; monique.ooms@wur.nl

Feed intake in newly weaned piglets is generally low, resulting in a poor performance postweaning. We investigated effects of pre- and postweaning enrichment and personality on feeding-related behaviours and growth of weaned piglets. Piglets were housed preweaning in barren 9 m^2 or in enriched (straw, peat, woodshavings) 18 m^2 pens with creep feed, and classified as high- or low-resisters using a backtest at day 15. At weaning (day 28), piglets were mixed (within preweaning treatment), with four animals per litter allocated to barren 3.2 m^2 and four to enriched 6.4 m^2 pens, and with two high-resisters and two low-resisters per pen. Piglets were observed at weaning using continuous focal sampling (n=128) and weighed at days 0 and 4 postweaning (n=256). Data were analyzed with mixed models; housing conditions were tested against pen. Preweaning enrichment reduced the latency to eat (14.2±2.8 vs. 23.2 ±2.9 min, $P<0.01$) and increased feed intake over the first 48h postweaning (0.53±0.04 vs. 0.45±0.05 kg/pig, $P<0.05$). Time spent eating was unaffected by housing conditions. Postweaning enrichment increased the latency to eat (25.3±3.1 vs. 12.1±2.5 min, $P<0.001$) and reduced time spent nosing the feeder (1.1±0.2 vs. 1.5±0.1%), but increased growth over the first four days (1.0±0.03 vs. 0.81±0.04 kg, $P<0.01$). Irrespective of housing conditions, high-resisters started to eat sooner than low-resisters (14.9±2.3 vs. 22.2±3.3 min, $P<0.05$), whereas low-resisters tended to nose the feeder more (1.5 ±0.2 vs. 1.2±0.1%, $P=0.10$). Backtest classification did not affect weaning weight, but high-resisters gained more weight until day 4 than low-resisters (0.96±0.04 vs. 0.85±0.04 kg, $P<0.06$). Preweaning enrichment enhanced feed intake, but postweaning enrichment and personality affected growth immediately after weaning.

Sexual behavior of Farahani ewes after estrus synchronization out of season

Moghaddam, Aliasghar[1], Kohram, Hamid[2] and Sohrabi, Ghadir[3], [1]Razi University, Department of Clinical Sciences, Faculty of Veterinary Medicine, Razi University, 67154-8-5414, Kermanshah, Iran, [2]Shahid Chamran University, Department of Clinical Sciences, Faculty of Veterinary Medicine, Shahid Chamran University, 5414, Ahvaz, Iran, [3]Khomeyn Azad University, Department of Animal Sciences, Faculty of Animal science, Khomeyn Azad University, 67154, Khomeyn, Iran; asgharmoghaddam2000@yahoo.co.uk

To compare the effect of the length (12 d vs. 6 d) of different progestagen treatments (cider vs. sponge) on reproductive efficiency, 24 ewes divided at random into four groups (6 in each), balanced for BCS, age and body weight. Groups C6 and S6 received intravaginal cider and sponge for 6 days, respectively, while groups C12 and S12 received cider and sponge for 12 days, respectively. At devices removal, all ewes received an i.m injection of 400 IU PMSG and then a ram was introduced into each paddock. The behavioral observation was conducted daily for five days to record proceptive behavior (soliciting, sniffing scrotum, head turning, anogenital sniffing and tail fanning) and receptive behavior (firm standing). ANOVA and Chi-square tests were used to analyze data. For significant test ($P<0.05$), the strength of association was calculated with eta squared. Frequency of proceptive behavior was increased at 12 h in groups C6, C12 and S6, whereas it increased at 6 h in group S12. Frequency of receptive behavior increased at 24 h in all groups. The perceptive behavior at 24 h was more in S6 than in C6 group ($P<0.05$). Estrus onset was 30 h for groups C6 and C12, and 29 h for groups S6 and S12 ($P>0.05$). Estrus duration was 18, 19, 19 and 20 h for groups S6, S12, C6 and C12 ($P>0.05$), respectively. A 100% pregnancy rate was obtained in all groups. Lambing and multiple birth rates were similar in C6 (166% and 40%), C12 (150% and 33.3%) and S12 (150% and 33.3%) which were more ($P<0.05$) than those in S6 (117% and 14.3%). In conclusion, the 6-day-cider or sponge–PMSG regimens could adequately replace the 12-day-cider or sponge-PMSG, out-of-season. Application of such a 6-day-cider or sponge-PMSG protocol is more cost effective with good results.

Social network analysis of behavioural interactions influencing the development of fin damage in Atlantic salmon (*Salmo salar*) held at different stocking densities

Cañon Jones, Hernán Alberto[1], Noble, Chris[2], Damsgård, Børge[2], Carter, Toby[3], Broom, Donald M.[1] and Pearce, Gareth P.[1], [1]University of Cambridge, Department of Veterinary Medicine, Madingley Road, CB3 0ES, Cambridge, United Kingdom, [2]Nofima Marine, Muninbakken 9-13, Breivika, Tromsø, Norway, [3]Anglia Ruskin University, Department of Life Sciences, East Road, CB1 1PT, Cambridge, United Kingdom; hac39@cam.ac.uk

The effect of high (HD, 30 kg/m^3) and low (LD, 8 kg/m^3) stocking densities on the occurrence of fin damage in Atlantic salmon was related to behavioural interactions using social network analysis. Four tanks (300 l) for HD and four for LD with 10 fish each were used. Dorsal fin damage was significantly higher in HD (15% vs. 2.5% of fish affected, $P<0.05$) with higher amounts of fin erosion in HD groups. Social networks based on aggressive interactions showed higher centrality (32.9% vs. 27.7%, $P<0.05$), clustering coefficient (0.43 vs. 0.41, $P<0.05$), in-degree centrality (35.2% vs. 34.0%, $P<0.05$), out-degree centrality (33.3% vs. 28.8%, $P<0.05$) and were less dense (64.2 vs. 73.9, $P<0.001$) in HD. High centralities and clustering coefficients indicated distinctive separation of fish within HD groups into initiators of aggression (out-degree four times higher than in-degree) and receivers of aggression (in-degree four times higher than out-degree). This separation of roles was seen only in HD groups where initiators had higher out-degree centrality (59.3% vs. 5.5%, $P<0.01$) while receivers showed high in-degree centrality (21.7% vs. 1.11%, $P<0.01$). Fish initiating aggressive interactions had less fin erosion (2.5% vs. 12.5% of fish, $P<0.05$), higher final weights (119.20 g vs. 105.00 g, $P<0.05$) and lengths (21.58 cm vs. 20.7 cm, $P<0.05$). HD groups had significantly less total aggression than LD groups (581.5 vs. 805.0 mean interactions/56 hours/group, $P<0.001$) and significantly higher biting events (5.25 vs. 1.75 mean interactions/56 hours/group, $P<0.05$) explaining the high fin damage in HD. Fish in LD groups had lower final weights, final lengths and body condition ($P<0.05$). High and low densities have differential detrimental effect in the welfare of farmed Atlantic salmon.

Changes in feeding behaviour of primiparous and multiparous cows during early lactation

Kindler, Annabell, Azizi, Osman, Julia, Dollinger and Kaufmann, Otto, Humboldt-Universität zu Berlin, Faculty of Agriculture and Horticulture, Department of Plant and Animal Sciences, Division of Animal Husbandry, Philippstrasse 13, 10115 Berlin, Germany; annabell.kindler@agrar.hu-berlin.de

Feed intake in high yielding dairy cows increases during the first month of lactation. This is affected by the individual feeding behaviour and influenced by factors like management practices or social interactions. Feeding behaviour is characterized by the number of visits at the feeder, the visit duration and the feeding rate. The aim of this research was to investigate general differences and changes over time of feeding behaviour between primiparous and multiparous cows. Therefore, daily feed intake and number and duration of visits from 70 German Holstein Frisian dairy cows (23 primiparous, 47 multiparous) have been analysed from the day after calving to 105th day of lactation (concentrating on the first two weeks). Both groups (primiparous and multiparous cows) showed a significant increase in daily feed intake from the first to the second week of lactation (primiparous: 31.94 ± 4.85 kg <40.33 ± 4.90 kg; multiparous: 37.72 ± 5.49 kg <49.53 ± 6.83 kg; $P \leq 0.001$). Focussing on changes in feeding behaviour, primiparous cows showed a significant increase in the visit duration d^{-1} from the first to the second week (2.45 ± 0.58 h <3.32 ± 0.84 h, $P \leq 0.001$) with a constant number of visits d^{-1}. In opposite, multiparous cows showed both: a significant increase in the visit duration d^{-1} (2.41 ±0.69 h <3.32 ± 1.04 h, $P \leq 0.001$) and a significant increase in the number of visits d^{-1} (27.28 ± 11.46 <33.53 ± 13.28, $P \leq 0.05$) from the first to the second week. The results indicate that there is a general difference in the feeding behaviour between primiparous and multiparous cows which should be considered by the management. The reasons are assumed to be social interactions between the animals (e.g. ranking) and adaptation processes of primiparous cows.

Which dairy calves are cross-sucked?

Laukkanen, Helja[1,2], Rushen, Jeffrey[1] and De Passille, Anne Marie[1], [1]Agriculture and Agri-Food Canada, Pacific Agri-Food Research Centre, Highway 7, Agassiz, BC V0M 1A0, Canada, [2]The University of Edinburgh, Royal (Dick) School of Veterinary Studies & Scottish Agricultural College, Roslin, Scotland, United Kingdom; Jeff.Rushen@agr.gc.ca

The factors that induce group-housed dairy calves to perform cross-sucking have been investigated but the duration of cross-sucking will also depend on the behaviour of the recipient. These social aspects of cross-sucking have not been examined. We examined which calves within a group were the recipients of cross-sucking, whether this was related to body weight or feeding behaviour, and whether cross-sucking served to displace calves from the feeders. Forty-five Holstein dairy calves housed in groups of 9 and fed by computer-controlled feeders were used. Calves were fed 6- 12l/d of milk and gradually weaned off milk either by d47 or d89. Cross-sucking was recorded continuously for 12 h/d on 2 days at approximately 37, 47, 79 and 89 days of age. Data on feeding behaviour were collected automatically by the feeders. 90% of calves were cross-sucked at least once but there were large differences between calves in the duration of time being cross-sucked. The top quartile of calves that were cross-sucked accounted for 68% of the total duration of cross-sucking. There was a moderate positive correlation across calves in the duration of being cross-sucked and the duration of cross-sucking another calf (r = 0.39, $P=0.008$). These calves had significantly higher bodyweights and spent more time in the milk feeders than the lowest quartile at most ages ($P<0.05$). Only a minority of cross-sucking occurred while the calf was actually in a feeder and in 80% of these cases, the receiver remained in the feeder. Cross-sucking rarely resulted in a calf being displaced from a feeder. We concluded that larger calves that spend more time in the feeders seem to be at a higher risk for being cross-sucked but that cross-sucking does not appear to be displace calves from the feeders.

Do laying hens of different coping styles show differences in coping with changes in social density and group size?

Koene, Paul, Wageningen University, Animal Sciences, Marijkeweg 40, 6709 PG Wageningen, Netherlands; paul.koene@wur.nl

High density and large group size have strong impacts on physiology and behaviour and consequently on health and welfare of farmed animals. Effects of social density and group size on vocalisation and behaviour of laying hens selected for high or low immune response on Sheep Red Blood Cells (SRBC-lines) were measured. Subsamples of 60 laying hens were pretested for differences in tonic immobility and open-field behaviour. They were tested repeatedly in 25 crossed combinations of group size (2, 4, 6, 8, and 10 birds) and density (1, 3.2, 10, 17.8 and 31.6 hens/m^2) during tests of 15 minutes. Using behavioural sampling, all vocalisations and behaviours were measured. Data reduction was done using factor analysis. With a polynomial GLM the relations between factor scores of vocalizations and overt behaviour, density, group size and SRBC-line were analyzed. Tonic immobility duration of the SRBC-lines (High 310 sec vs. Low 626 sec; $P<0.001$) may indicate differences in coping style. The four factors extracted from the test variables - with highest loadings of chasing, preen, ground pecking and contact call - were labelled Aggression, Comfort, Pecking and Calls. A significant negative linear relation between Comfort and social density was found (B=-1.74, $P<0.001$). Also a negative relation was found between Calls and group size (B=-0.62, $P<0.001$). No interactions between social density or group size and SRBC-line were found. Duncan post-hoc tests showed than the High line showed significant more Aggression (0.33 vs. -0.17), more Comfort (0.22 vs. -0.33), less Pecking (-0.37 vs. 0.48) and more Calls (0.39 vs. -0.26) than the Low line. Although the frequency of behaviours and vocalizations differ between SRBC-lines, the absence of interactions between SRBC-line and density or group size indicates that the SRBC-lines cope in the same way with increases in social density (less Comfort) and group size (less Calls).

Licking behaviour as social cohesion tool in cows on pasture

Pinheiro Machado, Thiago M.[1], Pinheiro Machado Filho, L. Carlos[1], Hötzel, Maria J.[1], Weickert, Lizzy C.S.G.[1], Volpato, Rodrigo M.[1] and Guidoni, Antonio L.[2], [1]LETA - Laboratório de Etologia Aplicada, Departamento de Zootecnia e Desenvolvimento Rural, Universidade Federal de Santa Catarina, Rod. Admar Gonzaga 1346 – Itacorubi., 88034-001, Florianópolis, SC, Brazil, [2]EMBRAPA, Suínos e Aves, BR 153, km 110, Vila Tamanduá, 89700-000, Concórdia, SC, Brazil; pinheiro@cca.ufsc.br

The role of allogrooming in cattle is still unclear, and it has been mostly related to physiological causes, hygiene and social submission. This study focused on the role of licking in dairy cattle as a socio-positive behaviour. Six Jersey herds, averaging 24.6±5 lactating cows on pasture under rotational grazing were investigated. The cows averaged 3.6±0.2 BCS (body condition score), and did not differ among herds. Each herd was observed for six consecutive days, between milkings, from 8am to 3pm, after four days of habituation. Every six minutes a scan of the herd was made and the behaviour of each individual cow was registered, as well as its closest neighbour. Licking and agonistic interactions were recorded as events, identifying executor and receptor of the behaviour. Sociometric matrixes were built for each herd. The percentage of cows licking and being licked during the 42h of observation was almost 100% for all herds. No correlation was found between executing and receiving licks, nor between licking or being licked with agonistic interactions ($P>0.05$). A high correlation ($R2=0,84$; $P<0.01$), however, was found between the number of aggressions of the dominant cow and the total aggressions in its herd. A positive ($P<0.0001$) correlation was found between licking and frequency of proximity of the pairs while grazing. Licking behaviour occurred mostly between 9 and 11am ($P<0.05$), coinciding with grazing bouts. In fact, the most common behaviour before and after licking was grazing ($P<0.05$). When not grazing, cows would be idling or ruminating, standing or lying. During the frame time observed, licking was a behaviour associated with grazing, but not with other behaviours. Grazing is an activity usually occurring in open areas, where animals are distracted selecting forages, and exposed to predation. In this situation licking behaviour might have evolved as a social cohesion tool for grazing herds.

Dynamic pattern, consequences and economic importance of tail-biting epidemics in fattening pigs

Sinisalo, Alina[1], Niemi, Jarkko[1] and Valros, Anna[2], [1]MTT Agrifood Research Finland, Economic Research, Luutnantintie 13, 00410 Helsinki, Finland, [2]University of Helsinki, Department of Production Animal Medicine, P.O. Box 57, 00014 University of Helsinki, Finland; jarkko.niemi@mtt.fi

Tail-biting in pigs is stress-induced multi-causal behavioural disorder causing lowered animal welfare and economic losses. Dynamic patterns of tail-biting outbreaks, consequences and economic importance of tail-biting in pig production were investigated based on individual records for 6812 pigs (boars 30.3%, females 29.9%, barrows 39.8%) from a Finnish progeny test farm. Non-parametric Mann-Whitney U-test and X^2 tests were used to study differences between the biters, the tail-biting victims and other pigs. Altogether 756 (11.1%) pigs were tail-biting victims and 33 (0.5%) were identified as biters. Boars were bitten most (14.6%) and females (11.3%) were bitten more than barrows (8.3%) ($P<0.001$). Boars were indicated as biters (0.8%) more often than females (0.5%) or barrows (0.2%) ($P=0.012$). Data contained both single-sex and double-sex pens. The percentage of tail-biting victims was 6.8% in barrow-only, 13.5% in female-only and 14.0% in boar-only pens. The average daily gain (from 34kg to 99kg) was 19 to 55g/d (depending on sex) lower for bitten than non-bitten pigs ($P<0.001$). Non-bitten pigs had thicker backfat and muscle than bitten pigs ($P<0.001$). Genetic differences between victims, biters and other pigs were quite small but further analysis is needed. Economic impacts and the development of tail-biting in the pen over time were estimated with a stochastic dynamic simulation model. When the probability of tail-biting outbreak in the pen increased above 40%, the prevalence of tail-biting started to increase rapidly. If the probability of an outbreak increased from 40% to 50%, losses per pig place increased from 3.0€/year to 8.7€/year. If the risk was reduced that much, a farm having 500 pigs was expected to save approximately 2,800€/year. These results suggest that immediate response to the tail-biting incident is important in preventing further economic damage.

The effect of box-size and having a companion on lying

Herbertsson, Sigtryggur Veigar[1] and Thorhallsdottir, Anna Gudrun[2], [1]Agricultural University of Iceland, dep. of Natural Resources, Hvanneyri, 311 Borgarnes, Iceland, [2]Agricultural University of Iceland, dep. of Environment, Hvanneyri, 311 Borgarnes, Iceland; annagudrun@lbhi.is

Limited work has been conducted on the effect of box size on the behavior of housed horses. The objective of the study was to analyze the effect of box size on the behavior of horses stabled individually or in pairs. 16 adult Icelandic horses, 8 geldings and 8 mares aged 4-13, were used in the experiment. All the horses were born and raised on the same farm were they belonged to the same group. The horses were randomly assigned to 4 different experimental groups, were the behavior of the same individual was observed, given different floor space, without a companion – (group 1: 3 and 6 m^2 without and 6 and 12 m^2 with a companion; group 2: 6 and 3 m^2 without and 12 and 6 m^2 with a companion; group 3: 4 and 8 m^2 without and 8 and 16 m^2 with a companion; group 4: 8 and 4 m^2 without and 16 and 8 m^2 with a companion). The experiment was divided into four experimental periods of 3 days each, with 6 days habituation periods between. Experimental periods were video recorded and recordings analysed for selected behaviors i.e. lying and eating behavior. No significant difference was between sexes regarding lying behavior. Sternal recumbency was negatively correlated with age ($P<0.001$). Lying behaviour was affected both by having a companion and by box size, as total sternal recumbency time ($P<0.01$) and length of lying bouts ($P<0.001$) increased with box size in the individual boxes but not in the pair boxes. When with a companion, the horses spent less time eating their ration than alone (83 vs. 144 min; $P<0.001$). The study shows that the smallest box sizes negatively affected lying behavior of the horses, both in total recumbency time and lying bout length and having a companion in the box resulted in less eating time. Both can have a negative effect on horse´s welfare.

What feeding rack design for goats kept in small groups?

Hillmann, Edna[1], Hilfiker, Sandra[1], Stauffacher, Markus[1], Aschwanden, Janine[2] and Keil, Nina Maria[2], [1]ETH Zurich, Institute of Plant, Animal and Agroecosystem Sciences, Universitaetsstrasse 2, 8092 Zurich, Switzerland, [2]Swiss Federal Veterinary Office, Centre for proper housing of ruminants and pigs, ART, Taenikon, 8356 Ettenhausen, Switzerland; edna-hillmann@ethz.ch

Especially for goats kept in small groups, as common in Switzerland, the design of the feeding area is delicate. The strict hierarchy among goats leads not only to agonistic interactions but bears the risk of low ranking animals not getting adequate access to feed. Aim of this study was to test the effect of partitions and head locks at the feeding rack in horned and hornless goats. With a total of 54 non-lactating dairy goats kept in eight groups (four horned, four hornless; six groups with 7, two groups with 6), four different types of feeding racks (partitions yes/no, head locks yes/no) were tested in a 2×2-factorial design. Each type was applied to each group for 5-6 weeks with an animal-feeding-place-ratio of 1:1. Agonistic and feeding behaviour was observed during the main feeding times. Data were analysed using generalised linear mixed-effects models with head lock, partition, presence of horns, dominance index and their 2-way-interactions as fixed effects. The effect of feeding rack design on agonistic and feeding behaviour strongly depended on rank and presence of horns. With head locks, low ranking ($F_{1,156}=74.55$, $P<0.001$) and horned goats ($F_{1,156}=16.83$, $P<0.001$) had a longer feeding duration. Feeding duration was further increased when partitions were present ($F_{1,156}=15.75$, $P<0.001$), especially in low ranking goats ($F_{1,156}=4.96$, $P=0.03$). Offensive aggression was only reduced when partitions were present additionally to head locks ($F_{1,156}=22.28$, $P<0.001$). Without head locks or partitions, goats showed 13.2 ± 0.3 aggressive interactions per feeding time. Partitions alone led to 16.2 ± 0.3 aggressions per goat, and head locks alone increased aggressions to 26.0 ± 0.5 per goat. Additional partitions reduced aggressions to 5.3 ± 0.1 per goat and feeding time. Thus, head locks enabled low ranking and horned goats to feed during the main feeding time. However, when using head locks, additional partitions are needed to prevent aggressions between neighbouring goats.

The influence of changing housing system on dairy cows' behaviour

Pavlenko, Anne[1], Kaart, Tanel[1], Poikalainen, Väino[1], Lidfors, Lena[2] and Aland, Andres[1],
[1]Estonian University of Life Sciences, Kreutzwaldi 1, 51014, Estonia, [2]Swedish University of
Agricultural Sciences, P.O. Box 234, Skara, SE-532 23, Sweden; anne.pavlenko@emu.ee

The aim of this study was to investigate dairy cows' behavioural changes during the first five weeks after transition to a new environmen. The study was performed on a new loose-housing cowshed where most of the cows were transferred from six tied housing cowsheds. Cows with serious health problems were not relocated. Behavioural observations were carried out in the first three days of the first, second and fifth week after transition to a loose housing system. The group of cows (n=105) was monitored during 18 ten minute periodsbetween morning and evening milkings. For each period the number of cows' showing studied behaviours was recorded. The GLIMMIX procedure with the SAS program was used to construct the logistic model to estimate the discrete effects of week, and additional external stress factors and linear trends among days within weeks, considering non-zero correlations between surveys made on the same day. Standing and moving behaviour were not different between study weeks but both vocalizing and aggressiveness were higher in the fifth week ($P<0.05$). Different external stress factors in the new environment, e.g. introduction of new animals into an already established group, new feeding system, exposure to unfamiliar people and group mates, were investigated further and found to increase standing, activity and frequency of vocalisation of the cows ($P<0.05$). It was concluded that the transition of dairy cows to a new environment had a significant influence on some behavioural patterns up to five weeks after relocation.

Aggression at mixing in growing pigs: the effect of weight, weight variation, and recent fight success

D'eath, Richard B. and Turner, Simon P., SAC, Animal Behaviour and Welfare, Edinburgh, EH9 3JG, United Kingdom; rick.death@sac.ac.uk

Aggression between pigs at mixing results in injuries and social stress, reducing welfare and production. 79 unisex groups of 15 pigs (n=1,185) were mixed at 71(±4.4) days of age (27.5±5.4 kg), each group contained five littermate subgroups of three pigs. Our aim was to determine the effect of weight and weight variation on aggression over 24hrs (two-sided fighting (4,899 occurrences) and one-sided bullying (4,490 occurrences)) and on skin lesion counts at 24hrs. The average weight of a group did not affect the frequency or duration of aggression, but heavy groups had more skin lesions (correlation r=0.267, P=0.017). Pigs in more uniform groups (low coefficient of variation for weight) spent longer fighting (r=-0.284, P=0.011) and had more skin lesions (r=-0.225, P=0.047). In ANOVA models blocking by group, heavier pigs within a group were involved in more fighting ($F_{(1,1182)}$=62.8, P<0.001) and bullying of other pigs ($F_{(1,1178)}$=41.41, P<0.001) and had more skin lesions ($F_{(1,1177)}$=6.27, P=0.012). In ANOVA models using data from individual interactions (blocking by group and pig) aggression between pigs with a large weight difference was shorter in duration (Fights, $F_{(1,3908)}$=3.96, P=0.047; Bullying, $F_{(1,3589)}$=5.76, P=0.016), fights involving heavier pigs lasted longer ($F_{(1,866)}$= 21.71, P<0.001), and heavier dyads were more likely to fight more than once ($F_{(1,3975)}$=6.55, P=0.011). To assess winner/loser effects, pigs which won at least one fight and lost at least one fight were identified (403 pigs; 2,687 fights) and ANOVA models included group and pig. When defeated, a pig had a longer latency until its next fight than when it won ($F_{(1,1932)}$=92.7, P<0.001). Following defeat, pigs were also less likely to initiate (χ^2_1=21.1, P<0.001) or to win (χ^2_1=81.0, P<0.001) their next fight. The variation in aggression explained by weight and winner/loser effects was significant but quite small. Nevertheless, producers should avoid mixing heavier pigs and similar-weight pigs if possible.

Do stallions affect the nature of interactions between their own harem members?

Granquist, Sandra M.[1], Sigurjonsdottir, Hrefna[2] and Thorhallsdottir, Anna Gudrun[3], [1]Institute of Freshwater Fisheries/The Icelandic Seal Center, Brekkugata 2, 530 Hvammstangi, Iceland, [2]The University of Iceland, Stakkahlíð, Reykjavík, Iceland, [3]The Agricultural University of Iceland, Hvanneyri, Borganes, Iceland; sandra@veidimal.is

Earlier researches indicate that stallions may affect the social behaviour in their harem, either by preventing interactions between harem members directly or indirectly by their presence. It has been suggested that this may lead to less interactions, unstable friendship bonds and less stable hierarchies between herd members in natural harems, compared to groups without stallions. In this study, we investigated interactions, friendship bonds and hierarchies in natural harems and compared to results described for similar groups without stallions. In addition, the effect of group stability on social behaviour was investigated by comparing temporary and permanent groups. 6 groups of horses were studied for altogether 525 hours (~80h/group). 4 of the 6 groups were permanently living together in a semi-feral herd. These groups (n=12-33) each contained 1 stallion, 8-23 mares, foals and 3-9 immature horses and were observed for 4 weeks. The other 2 groups were temporary breeding groups, consisting of 1 stallion, 27 resp. 32 unfamiliar mares and their foals. These groups were put together for approximately 6 weeks and where observed for the whole period. Interactions were recorded by all occurrence of some behaviour. The results show that stallions seldom intervene directly in interactions between harem members (0-0.05 times/h). However, significantly linear hierarchies were only developed in 3 of 6 groups (Landau's h´, $P<0.05$) and fewer friendship bonds were found (average 1-3, X^2–tests, $P<0.05$), compared to what have been described in groups without stallions. These results give some support to our prediction that stallion may indirectly suppress interactions between herd members. Group instability increased aggression rate (Kruskal-Wallis: $P=0.039$) and the most unstable group had significantly fewer friendship bonds compared to all the other groups (U-tests $P<0.01$). The results have significance for further research on social structure of mammals, and may be applied in management of domestic animals.

Relationship between fear of humans and amount of human contact in commercially reared turkeys

Botheras, Naomi A.[1], Pempek, Jessica A.[1], Enigk, Drew K.[1] and Hemsworth, Paul H.[2], [1]The Ohio State University, Department of Animal Sciences, 2029 Fyffe Court, Columbus, Ohio 43210, USA, [2]Animal Welfare Science Centre, University of Melbourne, Parkville 3010, Australia; botheras.1@osu.edu

Experimental and on-farm studies with broilers and layers show an association between the birds' level of fear of humans and human factors such as amount and type of behavioral interaction. Experimental studies with turkeys have found similar results. This study investigated the relationship between fear of humans in commercially-reared turkeys and the amount of human contact those birds received. Twelve barns, each housing 5,800 male turkeys, were visited. At 4, 8 and 12 weeks of age, a stroll test was conducted to assess fear of humans using the approach and avoidance behavioral responses of the turkeys to a stationary and moving novel human. A video camera was used to record the number of turkeys on-screen during moving and stationary phases of the test. Stockpeople recorded for one week prior to each visit the amount of time they spent in the barns each day. Correlations between the average number of turkeys close to the experimenter during stationary and moving phases of the test and the average amount of time stockpeople spent in the barns each day were determined. Stockpeople spent 39 ± 3.2 min/day (mean$\pm$SE) in the barn when the turkeys were 4 weeks old, increasing to 54 ± 6.6 and 67 ± 7.0 min/day at 8 and 12 weeks of age, respectively. There were no significant correlations between the amount of time stockpeople spent in the barn and the number of birds close to the novel human during either moving or stationary phases of the test at 4, 8 or 12 weeks of age (all $P>0.05$). In contrast to findings with other commercially-raised poultry species, the results suggest that amount of human contact is not a significant predictor of fear of humans in commercially-raised turkeys. It may be that the quality or type of human interaction is more important than the amount of time per se.

The effect of cage conditions on affective behaviour and responsiveness of SD rats

Turner, Patricia, Ovari, Jelena, Pinelli, Christopher, Cummins, Erin and Leri, Francesco, University of Guelph, Pathobiology, Ontario Veterinary College, Guelph, ON N1G 2W1, Canada; pvturner@uoguelph.ca

The effect of housing on rodent well-being has received increasing scrutiny. Most rodent studies evaluate different caging paradigms over short periods of time, which may not be realistic for typical colony settings. Inappropriate housing may lead to boredom, aggression, stereotypies and altered metabolism, which in turn may affect experimental responses. In this study, we evaluated the effect of single (n=8/sex), paired (n=16/sex), and enriched group (n=4/cage, 20/sex) housing on the behaviour and affective state of male and female SD rats over 5 months. Parameters included body weight, food consumption, monthly behavioural scoring, fecal corticoid levels and nociceptive responses, object recognition, and elevated plus maze testing, as well as mRNA expression of CRHR1 (pituitary) and CRH (hypothalamus). There were no differences in food consumption or body weight between housing paradigms within each sex. By study end, singly housed males were less active ($P<0.022$) and less well-groomed, with more porphyrin staining. Significant mild to moderate dominance-related ritualized aggression occurred in paired males ($P<0.017$), and less barbering and alopecia was noted in group housed animals. Group-housed animals used nylabone enrichment and engaged in more allogrooming and exploratory cage behaviours than single or paired rats, and significantly increased abnormal behaviours, such as bar-biting, were noted over time in single and paired female rats ($P<0.002$). Housing paradigm had no effect on elevated plus, object recognition or hotplate latencies; however, hotplate responses were significantly more consistent and faster for group-housed animals than single or paired rats. Blunting of dark phase fecal corticoid levels occurred in all groups over time with no consistent differences in total fecal corticoid levels. There were no differences in hypothalamic CRH mRNA expression for either sex; however, pituitary CRHR1 mRNA levels were significantly increased in single male rats ($P<0.007$) and paired female rats ($P<0.006$). The results of this study suggest that low density group housing of SD rats may be a preferred husbandry method resulting in more consistent physiologic responses and less stereotypic behaviours

Comparison of estrous behavior between two Iranian local sheep breeds (Zel and Dalagh)

Toghdory, A. H., Young researchers Club-Islamic Azad University-Gorgan Branch, Gorgan, Iran, 4914739975, Iran; Toghdory@yahoo.com

A total of 32 Zel and 32 Dalagh ewes, three–five years of age were randomly allocated in order to determine the estrous behavior of Iranian Zel and Dalagh ewes. Each group of ewes assigned in pens with three rams for thirty days. All of the rams in two groups were similar. Estrous behavior such as mountings and vulva sniffings of all ewes recorded in all time by electronic device and ewe fertility was determined. The result showed that Dalagh ewes showed the estrous behavior very weak and low than Zel ewes and for this reason apparent fertility was lower in Dalagh than Zel (53% vs. 82%). Dalagh ewes showed the estrous behavior in special time of the day like night but Zel ewes showed the estrous behavior in all the time of the day.

Socialising gilts to cope as young sows in grouphousing

Rasmussen, Hanne Midtgaard, Danish Agriculture and Food Council, Pig Research Centre, Axelborg, Axeltorv 3, 1609 Copenhagen V, Denmark; hmr@lf.dk

The introduction of young sows to life in a group of older sows is stressful to the animals. Social rank graduates access to resources as feeding stations and lying areas. Our focus is on young sows as they have low social rank position that later in life improves with parity number and body weight. The aim of an ongoing study is to provide gilts with experience to cope with aggression from animals larger than themselves. It is being investigated whether socialising gilts affects positively the young sows' use of lying areas and feeding station. It is also being investigated if there are fewer claw and leg injuries and an increase in longevity. Gilts are socialised at around 6 months of age by letting 30 gilts from a stable group mingle with a group of 120 gestating sows for two weeks. A gate is opened, so the pen of the gilts and the pen of the sows now form one common pen. The gilts remain access to their wellknown pen, but can choose to enter the sow area. And sows will enter the gilts area. Registrations of sows lying in the dunging area of the pen, and of sows that do not eat their daily ration of feed in the feeding station are carried out during the first two parities of the sows. It is investigated if socialising can reduce the proportion of animals not accessing these resources after introduction to the group. It is also investigated if the time span before all sows access lying area and feeding stations is shorter for socialised young sows. Additionally, leg and claw lesions are recorded as is the life span of the sows as a measure of a more gentle hierarchy formation in the gestation group. At the moment 116 gilts have been socialised, 150 gilts are in the control group, and a few individuals have completed first parity. The experiment is designed to have 800 sows in each group. Preliminary results will be available in August 2010, however no sows will have completed two parities.

Effects of foraging enrichment on collared peccary (*Mammalia, Tayassuidae*)

Nogueira, Selene S. C., Calazans, Stella G., Costa, Thaise S. O. and Nogueira-Filho, Sérgio L. G., Universidade Estadual de Santa Cruz, Laboratório de Etologia Aplicada, Rod. Ilheus Itabuna, km 16, Ilheus, Bahia, Brazil, 45662-900, Brazil; selene.nogueira@pq.cnpq.br

Captive breeding of collared peccary (*Pecari tajacu*) is on the increase in Neotropical countries. Previous studies have reported that peccaries increase agonistic interactions during feeding period. Therefore, environment quality improvements might help to decrease aggression and improve the captive conditions. Thus, we evaluated the effects of foraging enrichment on the occurrence of agonistic and exploratory behaviors in three groups of captive collared peccary (totaling 18 individuals). Following an ABAB experimental design, (A-traditional feeders/B-challenge feeders/A-traditional feeders/B-challenge feeders plus feeding randomization time), we tested the use of challenge feeders so that to get the food, the animal must open the feeder's door. Moreover, during enriched conditions we changed the feeding spot daily, filling just half of the feeders, trying to promote the animal´s searching behavior. We observed each individual during 5-min daily at feeding times, from 08:00h to 09:00h or from 16:00h to 17:00h, with the continuous-recording based on focal animal sampling sessions. In phase B2 we employed a random time schedule between 08:00h and 17:00h. Every treatment was carried out for two weeks. We conducted ANOVA with repeated measures followed by *post hoc* Tukey HSD tests, when appropriate. During the enrichment phases the animals showed a decrease on agonistic interactions ($F3, 42=6.99$; $P=0.001$) and an increase on exploratory behavior ($F3, 42=15.91$; $P<0.0001$). The food intake, measured by the difference between the amounts of food provided minus the remaining amount on the following day, decreased during the enriched conditions. As the animals did not show live weight change ($F3, 42=0.954$; $P=0.427$), these results suggest that challenge feeder usage produced less foodstuffs waste. We conclude that the challenge feeders employed in our study can improve peccaries' welfare by increasing animal's activities and diminishing fights during feeding.

Influence of the availability of enrichment material on the behaviour of finishing pigs
Zwicker, Bettina, Gygax, Lorenz, Wechsler, Beat and Weber, Roland, Centre for proper housing of ruminants and pigs, Swiss Federal Veterinary Office, Research Station Agroscope Reckenholz-Tänikon ART, 8356 Ettenhausen, Switzerland; bettina.zwicker@googlemail.com

To investigate the effect of availability of enrichment material on the behaviour of finishing pigs the number of straw racks offered in a pen was varied. 216 finishing pigs with undocked tails were housed in groups of 27 pigs each for 12 weeks. They were kept in pens and had access to a barren outdoor area both with partly slatted floors. Using a cross-over Latin square design the number of straw racks (one, three, six or eight) was changed in each group every 21 days. The racks were filled twice a day with cut straw. Video recordings were made for 16.5 hours on the 2nd and 18th day after changing the number of racks. The number of pigs showing exploratory behaviour towards straw in the racks (ER) or straw which dropped out of the racks on the lying area (ESL) as well as the number of pigs directing abnormal manipulative behaviour towards penmates (belly nosing, rooting on extremities or tail) was recorded by 10 minute scan sampling. Displacements from the racks were recorded continuously. Data were analysed using linear mixed-effects models. With more racks the proportion of pigs showing ER ($F_{3,21} = 8.95$; $P<0.01$) and ESL ($F_{3,21} = 3.54$; $P= 0.032$) was higher. Whereas ER increased steadily ESL remained constant with six or eight racks. The proportion of pigs directing manipulative behaviour to penmates was higher when six or eight racks were offered ($F_{3,21} = 6.15$; $P<0.01$). The number of racks had no effect on the number of displacements from the racks. Pig behaviour was not influenced by the number of days that had elapsed since changing the number of racks. In conclusion, our results indicate that for a group size of 27 fattening pigs 6 racks are most suitable to induce exploratory behaviour.

Effects of environmental enrichment on parturition site preference in dairy goats and behavior in goat kids Murciano-Granadina under confinement

Rosas-Trigueros, Alma Patricia, Otal Salaverri, Julio, Quiles Sotillo, Alberto and Carrizosa, Juan, University of Murcia, Animal Production, Facultad de Veterinaria, Campus de Espinardo, Murcia, 30100, Spain; mvz_rosas@yahoo.com.mx

Goats have been subjected to production schemes with escalating levels of intensity, which imply nullification of the possibility to manifest natural behaviors. The objectives of the present study were: (1) Evaluate the effects of environmental enrichment (EE) in goat behavior patterns to select a parturition site. (2) Evaluate goat kids behavior at EE conditions under confinement. Non-enriched and enriched were the two treatments randomly assigned to pregnant goats (n=21 each group) and goat kids, with four replicas (n=25 each). In both experiments, group 1, non-enriched, was maintained at normal conditions. Group 2, enriched, had EE (straw bales). The available area for parturition was divided in areas 'A' and 'B' (each area had 6.5 × 5 m), in group 2, area 'A' had EE. The goats that selected each area during labor were recorded. In goat kids, both groups had free access to a bounded area (BA)(6 × 5 m). In the enriched group the BA had EE. Observations of goat kids in the BA were carried out by video, during 3 representative periods of the day (period 1: 10:00-11:00 h, period 2: 14:00-15:00 h, period 3: 18:00-19:00 h) during 42 days. The results showed that the control group selected area 'B' which was further from the gate whereas enriched goats selected area 'A' during labor, although area 'A' was closer to the gate (14/21 vs. 5/21) (Chi-Square Test, $P<0.01$). The number of goat kids that were in BA was higher than non-enriched, the mean of the four replicas for the 1st, 2nd and 3rd period were: 0.8±1.8 vs. 2.4±1.9; 1.0±2.3 vs. 6.2±5.7; 0.8±2.2 vs. 6.3±6.8, (ANOVA test, $P<0.001$). Statistical analysis were carried out using the program SPSS 15.0.1 (SPSS, 2006). This study suggests that simple and low cost changes in the environment have significant effects in parturition site preference in pregnant goats and in goat kids behavior under confinement, improving animal welfare.

Litter provision in furnished cages for laying hens may induce pecking, scratching and dustbathing behaviours

Guinebretière, Maryse, Huonnic, Didier, Taktak, Amel and Michel, Virginie, AFSSA - Agence Française de Sécurité Sanitaire des Aliments, Unité d'Epidémiologie et Bien-Être en Aviculture et Cuniculture, BP 53 Route de Beaucemaine, 22440 Ploufragan, France; m.guinebretiere@afssa.fr

In the context of the EU Directive, we compared furnished cages (FC) in term of animal welfare. In the experiment, six treatments of six cages were compared in a 3×2 experimental trial: 3 group sizes of FC (20, 40 and 60 hens) with the same space per hen, with or without litter (feed) distributed automatically on a mat in the pecking and scratching area (PSA). During light period (7am to 11pm), hens were videoed during 2 days between 29 and 36 weeks of age (A1), and between 55 and 61 weeks of age (A2). Dustbathings were counted every 30 minutes in the PSA (A1&2), and every hour in each area of the cage (A2). Pecking or scratching behaviours were counted each hour in each area of the cage (only A2). Impact of group size and litter distribution and their interaction were evaluated on these behaviours (sum of scans on 2 days) with ANOVA after having checked applications conditions. Group size did not influence any of these behaviours. In PSA, significantly more dustbathings were observed in cages with litter distribution (A2): 20.8±11.1 compared to 15.7±6.8 in cages without litter (P=0.03, n=18). Elsewhere, sham dustbathings were observed but in lesser extent than in PSA; and litter distribution had no effect on their quantity. In PSA, litter distribution induced high increase of number of pecking or scratching behaviours: 40.7±19.8 compared to 9.6± 6.4 in cages without litter (P<0.001, n=18). Elsewhere, litter distribution had no effect on the quantity of pecking and scratching behaviours. Whatever the group size, treatments with litter (feed) distribution in the PSA, every hour in our case, enabled hens to dustbath and explored more by pecking and scratching, increasing their welfare. However, sham dustbathings still exist even in presence of litter. Thus further research on PSA (surface, location, type and quantity of litter to distribute) is necessary.

Towards long(er) pig tails: new strategy to solve animal welfare problems
Bracke, Marc B.M., Wageningen UR Livestock Research, Edelhertweg 15, 8200 AB Lelystad, Netherlands; marc.bracke@wur.nl

Despite a considerable scientific knowledgebase, many animal-welfare problems remain. Intelligent Natural Design (IND) is a promising approach designed to reduce long-lasting multifactorial welfare problems such as tail docking and tail biting in pigs. IND combines the advantages of applying scientific knowledge and human intelligence with one of the most intriguing ways to solve complex design problems, namely natural selection. In order to reduce routine tail-docking in intensively-farmed pigs conditions for selection and evolution can be created artificially. A first generation of potential management strategies to deal with tail biting will be implemented in a selected number of farms informed by current scientific knowledge. Farmers may, for example, be implementing different enrichment materials in pens with undocked or partly docked pigs (which is in compliance with EC directive 2001/93 which prescribes raising undocked pigs to verify the need for routine tail docking). Selection of the best ('fittest') strategies from the first generation, e.g. as indicated by lowest levels of tail biting, will provide the starting point to implement the next generation comprising a new batch of farms or pigs adopting slightly modified replicates of previously best management strategies. Solutions can evolve in diverging directions, like species. To support the required information-exchange a database is built to assist farmers, extension and scientists to monitor and direct the IND process in preferred directions. Market opportunities, reducing existing problems with tail biting, meeting EU regulations, public concern, division of labour and ethical room for innovative manoeuvre should motivate farmers to participate in this process to 'grow back' the pig's tail. The IND methodology enhances but cannot promise actual solutions for welfare. Nevertheless, the point that natural selection may be used to improve welfare provides an interesting extension to the common view that the performance of natural behaviour is important for animal welfare.

Dog-human communication factors that affect the success of dog training

Browne, Clare M., Starkey, Nicola J., Foster, Mary T. and McEwan, James, University of Waikato, Department of Psychology, Private Bag 3105, Hamilton 3240, New Zealand; clare_browne@yahoo.co.nz

Research on dog-human communication shows dogs are extremely responsive to human cues, such as vocalisations, pointing, and eye gazing. Because dogs are so receptive to these cues, it is reasonable to assume that the quality of dog-human communication has an impact on the efficacy of dog training. The purpose of this study was to examine which cues dog owners and trainers believe are important during training. Four dog owners and five trainers were interviewed to examine their views of the importance of different types of human-given cues used during dog training. They were asked questions about dog-human communication during training activities, and also asked to rank the possible cues. The data showed high agreement amongst participants' rankings for proximity, volume of voice, and sudden movements. Proximity was considered reasonably important by most participants, while volume of voice was agreed to be of less importance. Sudden movements weren't considered very important. There was less agreement about tone of voice and hand gestures; however, tone of voice was considered more important than hand gestures. There was little agreement across participants on the importance of body position or pronunciation. A Friedman's test was done on eye contact, hand gestures, arm movements, tone of voice, and pronunciation, as the other cues had missing values. This was significant (c^2=12.62, df=4, P=0.01) showing that the rankings differed across these cues. Thus, the results indicate that while there was some agreement amongst participants on the importance of some cues during dog training, agreement wasn't consistent across all cues. In spite of this, there were some differences between the rankings. Tone of voice was considered the most important cue overall and sudden movements considered the least important. These are preliminary results. Of interest is the effectiveness of these cues in dog training.

Identification of behavioural QTLs in the chicken

Wright, Dominic[1], Jensen, Per[1] and Andersson, Leif[2], [1]Linköping Universitet, IFM biology, Department of Zoology, Linköping, 58183, Sweden, [2]Uppsala Universitet, IMBIM, BMC, Uppsala, 75124, Sweden; domwright@gmail.com

The domestication process can elicit a wide range of changes, acting on behavioural, morphological and life history characteristics (including bodyweight, bone density, anxiety behaviour and fecundity), to produce massive genetic changes in a species. The differences between a population of wild red junglefowl and a domesticated strain of White Leghorn chickens has been used as the basis of a QTL study to identify the genetic architecture of anxiety-related fear avoidance behaviour. The F2 intercross of these strains identified 12 significant or suggestive QTL (with LOD scores ranging from 2.9 to 5.4) affecting tonic immobility, latency to inspect a novel object, open field activity and restraint phenotypes on chromosomes 1, 2, 3, 4, 6, 15, 18 and 27. To further narrow down the regions containing these QTLs, an eight generation intercross has now been performed, comprising of 560 birds, and phenotyped for tonic immobility, novel object inspection and open field activity. Each of these tests were performed twice per animal, and was shown to have significant levels of repeatability (correlation coefficients of between 0.30 and 0.65 for various characters associated with the tests). This indicates that the maximum possible heritability for these traits is high, as the repeatability of a trait is also its maximum possible heritability, and the potential to identify candidate genes in these regions is now possible.

Outdoor use of broilers differing in growth intensity

Hörning, Bernhard[1], Trei, Gerriet[1] and Sabine, Gebhardt-Henrich[2], [1]University of Applied Sciences Eberswalde, Department of Organic Livestock Farming, Friedrich-Ebert-Str. 28, D-16225 Eberswalde, Germany, [2]Federal Veterinary Office, Centre for proper housing of poultry and rabbits, Burgerweg 22, 3052 Zollikofen, Switzerland; bhoerning@fh-eberswalde.de

According to a EU regulation, organic broilers must have access to outside runs and must originate from slow growing strains unless a minimum slaughter age of 81 days is given. Aim of the study was to compare outdoor use of seven broilers strains: Brahma and Cochin, two pure breeds (20-25 g daily gain), Sasso and Kabir (30-35 g), Olandia and Hubbard (40-45 g), Ross (60 g). Altogether, 30 groups of 50 broilers were kept in two consecutive fattening periods, the first being in summer, the second in autumn. Outdoor use was recorded using direct observations in the summer period. The number of broilers outside per group was counted at 12 days each at 5 times. An RFID technology system was used in 6 groups which allowed the recording of individual outdoor use. Only 2.0% of all broilers of a group were recorded outside during direct observations. 83% of these animals were found in the fourth nearest the poultry house. The average use for Cochin, Brahma, Kabir, Sasso, Olandia, Hubbard and Ross was 4.38, 1.09, 2.45, 1.95, 0.61, 1.54 and 0.04%. Ross broilers used the runs less than Cochin, Brahma, Kabir, Sasso and Hubbard ($P=0.001, 0.025, 0.001, 0.001, 0.012$). According to the automatic system, a total sum of 20102 outdoor uses was recorded at 70 days (287.2 per day). 60% of all individuals used the runs in the summer period and 35% in autumn. Broilers used the outside run on average 5.7 times per day in summer and 9.1 in autumn. Mean duration was 25.7 min. in summer and 10.1 min. in autumn. Mean temperature was higher in summer. Number of animals in the run per day correlated with duration per animal and day ($rs = 0.660$, Spearman rho). In conclusion, fast growing broilers used the outdoor run least.

Causal factors of body hygiene in dairy cows and its relationship to milk somatic cell count

Paranhos Da Costa, Mateus J.R.[1] and Sant'anna, Aline Cristina[2], [1]UNESP, Depto Zootecnia, Via de Acesso Prof. Paulo Donato Castellane, s/n Depto Zootecnia Prédio 2, 14887-900, Brazil, [2]UNESP, Programa de Pós Graduação em Genética e Melhoramento Animal, Via de Acesso Prof. Paulo Donato Castellane, s/n Depto Zootecnia Prédio 2, 14887-900, Brazil; mpcosta@fcav.unesp.br

The aim of this study was to describe variation in dairy cow body hygiene conditions over time, and to assess the relationship between body hygiene and cow behaviour, particularly that related to space use and thermoregulation, and somatic cell count (SCC). Cow body hygiene was assessed in 3,554 observations of 545 lactating cows, for nine months in two Brazilian commercial farms. The body hygiene was recorded on a four point cleanliness score: very clean, clean, dirty and very dirty. Behavioural categories considered were: posture (standing and lying), spatial position (under or not shade, on mud or dry surface) and activities (rumination, feeding and locomotion). At least 46 dairy cows per month were observed in four months, two in the dry season and two in the rainy season. The SCC was assessed in milk sample of 404 cows (2,218 samples). Individual differences were found in the cow body cleanliness score, and this also varied according to season ($X^2 = 3.17$, $P<0.01$). More than fifty percent of the cows were consistent in the body cleanliness score (45.86% consistently clean and 9.76% consistently dirty). The body cleanliness score was associated with cow behaviour, as follows: 1) clean cows tended to spend less time on mud surfaces (Kruskall-Wallis: H= 5.38, $P=0.06$); 2) very dirty cows lay down more frequently under shade (Kruskall-Wallis: H=10.61, $P<0.05$); 3) very dirty cows spent more time under shade (Kruskall-Wallis: H= 10.63, $P<0.05$). The body cleanliness score affected SCC (F = 4.85; df = 3; $P<0.01$), as very clean cows had lower SCC (4.60±2.33 somatic cell linear score) than the others (clean= 4.77±2.37, dirty= 4.97±2.28 and very dirty= 4.94±2.34). Based on these results we concluded that the state of hygiene of a cow's body depends on many factors such as behaviour, climate and management. However, in spite of this complexity, the cow's cleanliness score can be used for risk assessment of udder health problems.

Evaluation of on-farm veal calves' responses to humans: influence of group size on the outcomes of two tests

Leruste, Helene[1], Lensink, Joop[1], Bokkers, Eddie[2], Heutink, Leonie[3], Brscic, Marta[4], Cozzi, Giulio[4] and Van Reenen, Kees[3], [1]Groupe ISA, 48 boulevard Vauban, 59046 Lille, France, [2]WUR, Animal Production Systems, P.O. Box 338, 6700 AH Wageningen, Netherlands, [3]WUR, Livestock Research, P.O. Box 65, 6200 AB Lelystad, Netherlands, [4]University of Padova, Department of Animal Science, Viale dell'Università 16, 35020 Legnaro, Italy; h.leruste@isa-lille.fr

On-farm monitoring of animal welfare involves, amongst others, the assessment of animal's responses towards humans. The objective of this study was to investigate, in two different housing systems with different group sizes, if the voluntary approach of veal calves towards humans is associated with their response to being approached by a human. These tests, supposed to reflect fear of humans, were performed on 182 farms with small groups of <10 calves and 34 farms with large groups of >20 calves, aged 11-18 weeks. The mean latency (max 180s) for the first 5 calves to touch an immobile unfamiliar human standing at their pen's fence was recorded in 10 pens per farm (small groups) or all pens (large groups). The response of individual calves to being approached by an unfamiliar human inside the pen was assessed on a 5-point scale: 0 (no eye contact) to 4 (snout can be touched). On each farm, responses were obtained in 100 randomly chosen calves previously tested for the voluntary approach of the human. Analyses were performed at farm level. Mean latency (+SE) for voluntary touching the human was 59.7±1.3s without differences between housing systems. Mean escape score was lower in small groups than in large groups (1.7±0.1 vs 2.2±0.1; $P<0.01$), with more calves scored 0 (5±0.6% vs 2±0.6%; $P<0.01$) and less calves scored 4 (14±0.9% vs 28±2.2%; $P<0.01$). Among farms with small groups, approach latency was negatively correlated with average escape score and proportion of calves scored 4 ($r_s=-0.25$ and $r_s=-0.24$; $P<0.01$) but not significantly related to the proportion of calves scored 0. Among farms with large groups, no significant correlations were found between the two tests. These results suggest that calves housed in systems of different group size respond differently to the escape test. The two tests are weakly correlated and only when performed on calves housed in small groups. More work is needed to develop new tests that are comparable between housing systems.

Study of feed tossing behavior in cows using a wire cable emplaced in front of manger

Farivar, Fariba and Tatar, Ahmad, Gorgan University of Agricultural Sciences and Natural Resources, Dept. of Animal Sciences, Basij Square, 336 Gorgan, Iran; a_tatar2002@yahoo.com

Feed tossing behavior of cows eating from a manger leads to feed waste. This experiment was conducted to evaluate a new method for decreasing feed tossing by dairy cows. 45 lactating dairy cows were randomly allocated to three treatments: two groups had a cable in front of their manger (located 40 cm away from the feeding gate and 80 or 70 cm above the standing floor of cows), the other group had no cable above their manger (control). One week after fixing the cables, feed tossing behaviour of cows was observed and recorded during morning feeding of a total mixed ration for 45 min per day for 10 days by direct observation. Backward and forward tossing behaviours were recorded for each group and means were compared between groups using analysis of variance. There was a highly significant difference between groups in both backward and forward tossing behaviours ($P<0.01$). Both groups with cables above their manger showed significantly ($P<0.01$) less feed tossing behaviour (22 and 20 vs. 52 times backward feed tossing and 9.2 and 4.2 vs. 25.4 times forward feed tossing for 80 cm and 70 cm cabled treatments and control respectively), but height of cable (80 or 70 cm) had no significant ($P>0.05$) effect on these behaviours. This experiment showed that feed tossing behaviour can be decreased using a wire cable in front of the manger.

Effects of spatial disposal of grazing conditioners in the use of space by beef cattle: a case study

Pascoa, Adriano Gomes, Lima, Victor Abreu, Ferrarini, Carla and Paranhos Da Costa, Mateus José Rodrigues, UNESP, Depto Zootecnia, Via de Acesso Prof. Paulo Donato castelane, s/n Depto Zootecnia Prédio 2, 14887-900, Brazil; agpascoa@uol.com.br

The variation in the resources available in a pasture promotes the use of the area in a non-uniform way by the cattle. The aim of this study was to analyze the effects of spatial arrangement of the conditioning of grazing behavior of cattle with the use of a Geographic Information System (IDRISI) and the implication of these features in the space use by animals. The study was conducted in Uberaba, Brazil from January to March 2009. Three pastures were used which were differently in topography, provisions of resources (such as shading and watering), and animal categories (cows with calves, heifers and bulls). The pastures were geo-referenced and the animals were monitored with the use of GPS collars, placed on two animals in the pasture with cows and one animal in the pasture with the heifers and bulls. The collars were programmed to record the position of animals every minute for 15 days. The data were analyzed in GIS software. The results showed that in 66% of the time the animals were located on land sloping up to 5. The animals remained at an average distance of 22.95 ± 6.52 meters of the cattle tracks. There was significant difference between the average distances that animals kept from water in the three paddocks (ANOVA, $F2,164 = 38.75$, $P<0.01$), whereas the cows with the calves remained closer to water sources than heifers, which remained closer than the bulls. These results suggest that cattle, regardless of their category, prefer to be at low land slope and near to the tracks, a strategy that can work as energy saving. The variation in the position of animals in relation to water sources suggests that each animal category has different needs of this resource.

Influence of the slope of nest floor on nest choice of laying hens

Stämpfli, Karin, Buchwalder, Theres, Roth, Beatrice A. and Fröhlich, Ernst K. F., Centre for proper housing: poultry and rabbits, FVO, Burgerweg 22, 3052 Zollikofen, Switzerland; karin.stämpfli@bvet.admin.ch

Group nests in alternative housing systems for laying hens fulfil mostly a hen's need for seclusion and protection. Although nests are built according to welfare guidelines, there are differences between nest types such as floor slope which could have an impact on egg laying behaviour. Floor inclination has to be designed in such way that eggs roll away without breaking and hens feel comfortable to lay their eggs. In commercial nests the slope is usually between 12%-18%. The aim of this study was to investigate the effect of floor inclination on the hens nest preference and laying behaviour. We predict that hens prefer nests with a low-pitched floor because of evolutionary and comfort reasons. Eight pens each with 17-18 white laying hens (LSL) were equipped with two roll-away nests (0.54m^2) with different floor slopes (12% and 18%). The floors were covered with mats of rubber pimples. The laying behaviour was recorded for two days. Data were analysed with a repeated measures ANOVA (NCSS). More hens were counted in nests with 12%-slope (12%: 2.96±0.25; 18%: 1.99±0.25; P=0.027). The number of eggs tended to be higher in nests with 12%-slope (12%: 9.81±1.3; 18%: 6.91±1.3; P=0.160). More sitting periods (12%: 44.00±3.5; 18%: 24.63±3.5; P=0.006) were recorded with 12%-slope, whereas the mean duration of sitting periods tended to be higher in nests with 18%-slope (12%: 757s±45.2; 18%: 929s±60.7; P=0.074). The higher number of hens counted and the light trend for higher egg number in the 12%-nest indicate that hens preferred this nest. This is supported by the higher amount of sitting periods found in these nests. The longer duration of sitting periods in 18%-nests could be an indication of feeling uncomfortable to stand in the nest. We recommend for commercial egg production to offer group nests with a floor slope as low-pitched as possible.

Behavioural and emotional responses in pet dogs during performance of an obedience task in the absence of the owner

Torres-Pereira, Carla and Broom, Donald M., University of Cambridge, Centre for Animal Welfare & Anthrozoology, Department of Veterinary Medicine, Madingley Road, Cambridge CB3 0ES, United Kingdom; cmct2@cam.ac.uk

The purpose of this study was to characterise behavioural and emotional responses in dogs during performance of an obedience task in the owners' absence, and see the impact of using different objects on obedience to the task. 32 pet dogs with knowledge of the command 'Leave it!' were selected. Each animal performed two obedience tasks associated with the command 'Leave it!', following a within-subject and between-subject design. Obedience to the command and behaviours displayed while the owner was away and when he/she returned were video recorded and later observed. The owner placed an object (Toy or Food) on the table, said 'Leave it!', walked out of the room and closed the door. After 30 seconds, he/she returned, either Looking at the dog or being Indifferent. Binary data from behaviour were analysed in GenStat using GLM. Deviance associated with variate Taking Object is 63.04 on 63 d.f. from 64 observations. It suggests a highly significant difference between the two objects, Toy and Food ($P<0.001$). The model estimated that probability of Food being taken by the dogs is 0.41 whereas for Toy is 0.16. Probability of Wagging Tail is 0.53 for Food, 0.40 for Toy ($P<0.001$). This behaviour was displayed towards the owner to ask for the object on the table. The dogs obeyed (left object alone) in 33 of the 64 observations, did not take object in 46, took object in 18, took food in 13 and took toy in 5 observations. Food was taken more times; there was a higher probability of a dog taking Food than Toy and of wagging the tail to ask for Food than for Toy. Food showed higher incentive value and disobedience was higher for Food. Obedience to the command 'Leave it'! in the owner's absence depended on the object used in the obedience task.

Effects of positive versus negative prenatal handling on emotional reactivity in lambs
Coulon, Marjorie, Hild, Sophie, Schröer, Anne and Zanella, Adroaldo Jose, Norwegian School of Veterinary Science, Department of Production Animal Clinical Sciences, P.O. Box 8146 Dep., 0033 OSLO, Norway; marjoriecoulon82@yahoo.fr

In farm animals, stressful events during pregnancy can affect the offspring and their adaptation to the environment. The effects may vary with intensity of the stressor and by how aversively it is perceived by the dams. Negative human handling is stressful and has impact on welfare outcomes, while positive relationships can reduce fear of humans. Our aim was to study the effect of negative or positive handling (loud shouting-abrupt movement versus calm behaviour-soft talking), imposed twice daily for 10 min, on 12 multiparous and 12 primiparous Norwegian-dala ewes, during the last five weeks of pregnancy, on the behaviour of their lambs. Lambs prenatally negatively treated (PN: N=12) and positively treated (PP: N=11) were individually tested at 4 weeks of age. Lamb behaviour was recorded during a 'human' test, composed of one isolation phase and one phase with an unfamiliar human present (4 min each) and a startle test consisting of the sudden opening of an umbrella followed by 5 min behaviour recording. During the human test, no significant effects of treatment, parity or interaction between treatment and parity were found on overall recorded behaviours (ANOVA). However, PN lambs vocalized and explored the environment less and were slower to approach the human zone when the human was present (Post-hoc comparisons: $P<0.001$, $P<0.001$, $P<0.001$). Fear of humans may have been transmitted to the young by the dam in PN lambs in the post-natal period. In the startle test, we observed no effect of treatment but lambs from primiparous ewes spent more time in the umbrella zone, were less active and tended to vocalize less than lambs from multiparous animals (ANOVA: $P<0.01$, $P=0.017$, $P=0.064$). These changes in behaviour showed by lambs from primiparous ewes may reflect a lower reactivity in a stressful situation. This could help the animals cope better with stress but be maladaptive in natural conditions where a higher reactivity decreases risks of predation.

The ontogeny of mother-young acoustic recognition in goats

Briefer, Elodie and Mcelligott, Alan, Queen Mary University of London, School of Biological and Chemical Sciences, Mile End Road, London E1 4NS, United Kingdom; e.briefer@qmul.ac.uk

Parent-offspring recognition is essential for the reproductive fitness of parents when parental investment is high and for survival of the young. So-called hider young spend most of their time hidden in vegetation to avoid detection by predators. Recognition studies in ungulate are sparse and need further investigations with species that vary in their ecology. We investigated mother-offspring recognition in goats (Capra hircus), a hider species in which an exclusive bonding to the neonate by the mother is developed within a few hours postpartum. We examined both the coding and the decoding process of individual identity between mother and offspring. Bleats of females and of kids produced during their first 7 days and at 5 weeks old were analysed to determine individual distinctiveness and to find potential individual signatures. Mutual recognition between mothers and kids was tested using playback experiments during weeks 1 and 5 after birth. Our results indicate mutual but asymetrical vocal recognition, both at week 1 and week 5. Kids responded strongly to both their own mother calls and familiar female calls, with a slightly shorter response to their own mother calls (Friedman ANOVA, N=15: latency of response, $F_{2,11}=9.43$, $P<0.05$; other response measures: NS). Adult females responded a lot more to their own kid calls compared to a familiar kid calls (Friedman ANOVA, N=9: $P<0.05$ for all response measures except latency to look toward the loudspeaker, $F_{3,5}=5.01$, $P=0.17$). These results are contrary to predictions for hider species, in which a unidirectional recognition of the mother by the young is expected. However, they lend support to the hypothesis that selection pressures on recognition behaviour can be distinct for mothers and young, with young responding more actively than mothers and making more recognition errors, indicating a higher level of motivation to reunite.

Changes in spectral baseline EEG in cattle during conditions requiring varying levels of primary visual processing

Drnec, Kim, Rietschel, Jeremy and Stricklin, W.R., University of Maryland, Cognitive and Neuro Science, College Park, 20742, Maryland, USA; kdrnec@umd.edu

Electroencephalography (EEG) has been successfully employed to objectively study cognitive processes in humans, notably pain. Animal research using EEG has been primarily for biomedical research rather than understanding species-specific cerebral-cortical dynamics that may underlie respective cognitive processes. Although cattle experience painful procedures, little is known about their cognitive processes, specifically how they process nociceptive stimuli. Our research goal is to develop an objective marker of nociception in cattle. As a first step reported herein, we recorded EEG using conscious, quietly standing cows to compare with human EEG results. The purpose of this initial information was to determine if cows exhibit shifts in spectral power that are consistent with those observed in humans, i.e., increased alpha power (8-12 Hz) during darkened room compared to lighted room. EEG were collected from cows (n=6) during counterbalanced conditions requiring varying levels of primary visual processing; specifically, exposure to natural light (3 candelas) and exposure to a darkened (0.4 candela) room. These data were Fourier transformed and normative frequency bands subjected to paired t-tests. In an attempt to identify species- specific functional frequencies, the spectral power estimates were subjected to Principal Component Analysis (PCA). The results demonstrated that delta power (1-4 Hz) increased significantly ($P<0.01$) during the darkened room as compared to the natural light condition, which is directionally consistent with the human literature. Moreover, multiple PCA factors representing potentially functionally distinct cognitive processes were significantly different ($P<0.05$) between the two conditions. These initial results implicate EEG as a viable metric to assess cognitive processes in cattle. Should this research lead to an objective methodology for identifying pain in animals as it has in humans, the methodology could have considerable importance to animal welfare.

Quality prevails over identity in the sexually selected vocalisations of an ageing mammal

Mcelligott, Alan G.[1], Vannoni, Elisabetta[2] and Briefer, Elodie[1], [1]Queen Mary, University of London, School of Biological and Chemical Sciences, Mild End Rd, London, E1 4NS, United Kingdom, [2]Universität Zürich, Anatomisches Institut, Winterthurerstrasse 190, 8057 Zürich, Switzerland; a.g.mcelligott@qmul.ac.uk

Male sexually selected vocalisations contain both individuality and quality (status) cues that are important for intra- and intersexual communication. As individuality is a fixed feature whereas male phenotypic quality changes with age, individuality and quality cues may be subjected to different selection pressures over time. We investigated this in fallow deer, in which some acoustic parameters of vocalisations code for both individuality and quality. We carried out a longitudinal analysis by investigating groan individuality and the effects of age and dominance status on the acoustic structure of groans, of the same males recorded during consecutive years. Both age- and dominance-related acoustic parameters contributed to individuality. There was a significant positive linear relationship between age and all measures of the fundamental frequency (F0) contour ($F0_{max}$, $F0_{mean}$ and $F0_{min}$), with males producing higher-pitched groans when older. This relationship was strongest for $F0_{min}$ (mean $F0_{min}$: 5 years old, N = 3, 19.39 ± 0.53 Hz; 6 years old, N = 12, 20.71 ± 0.50 Hz; 7 years old, N = 12, 22.05 ± 0.54; 8 years old, N = 4, 23.26 ± 0.98; $F_{1,16}$ = 42.8, $P<0.0001$). Male dominance changed with age, inducing a change in related vocal parameters and thus, a modification of vocal cues to male individuality between years. Dominance rank had a significant effect on $F0_{max}$, with males having lower-pitched groans when higher-ranking ($F_{1,12}$ = 6.15, $P<0.03$). Fallow deer vocalisations are honest signals of quality that are not fixed over time but change dynamically according to male status. As they are more reliable cues to quality than to individuality, they may not be used by conspecifics to recognize a given male from one year to another, but potentially used to assess male quality during each breeding season.

Emotional brain response to grooming is stronger in sheep with a negative mood
Gygax, L[1], Muehlemann, T[1,2], Reefmann, N[1,3,4], Wechsler, B[1] and Wolf, M[2], [1]Centre for Proper Housing of Ruminants and Pigs, Agroscope Res Stat ART, 8356 Ettenhausen, Switzerland, [2]USZ, Biomed Opt Res Lab, Frauenklinikstr 10, 8091 Zurich, Switzerland, [3]Behavior Biology, Badestr 13, 48149 Münster, Germany, [4]Ethology & Animal Welfare, P.O. Box 7068, 750 07 Uppsala, Sweden; lorenz.gygax@art.admin.ch

Welfare is reflected by the affective state of an animal which consists of both short-term emotions and long-term mood. Because affective states are generated and processed by the brain, we applied non-invasive functional near-infrared spectroscopy (fNIRS) to measure haemodynamic changes caused by neuronal activation. We investigated the interplay between positive and negative mood presumably induced by enriched and barren/unpredictable housing conditions, respectively, and short-term emotions. The study was carried out with nine freely moving sheep and short-term emotions were elicited by grooming by a human experimenter (previously been shown to be of positive valence) and compared to non-grooming periods. We modelled the time course of the haemodynamic brain response, i.e. the concentration changes of oxyhaemoglobin (O_2Hb), using mixed-models accounting for the individual sheep. We simultaneously tested the influence of mood and brain location on the time course. The changes in O_2Hb concentrations were temporally consistent with grooming onset and end and depended significantly on the subjects' mood and on the longitudinal localisation (caudal vs. rostral) of the response in the cerebral cortex (interaction time × mood × longitudinal localisation: $F_{17,5260} = 3.60$, $P<0.0001$): the O_2Hb concentration in the caudal cortex decreased markedly for the animals in a negative mood (AUC mean±SE: -51±4 µmol/l), at the same time as it increased in the rostral regions (35±16). By contrast, O_2Hb decreased less sharply for the animals in a positive mood in both the caudal (-22±9) and rostral cortices, but in a somewhat delayed manner in the rostral cortex (-9±23). This pattern is consistent with the notion that an emotional activation occurred in the frontal cortex of the animals and was stronger in animals in a presumably more negative mood. Assuming that this activation reflects emotional reactions, this finding contradicts the assumption that negative mood generally taints reactions to emotional stimuli.

Are ewes affected by castration and tail-docking of their lambs?

Edgar, Joanne L, Clark, Corinna C A, Paul, Elizabeth S and Nicol, Christine J, University of Bristol, Clinical Veterinary Science, Langford House, Langford, Bristol, BS40 5DU, United Kingdom; j.edgar@bristol.ac.uk

Castration and tail-docking (C/TD) are routine husbandry procedures that cause pain and stress in lambs. However, there have been few studies on the effects these have on mother ewes. The aim of the current study was to (a) assess whether ewes were affected by C/TD of their lambs and (b) assess whether the response was affected by the ewe's own neonatal experience. At 72-96 hours of age 20 Suffolk X Mule ewes were randomly assigned to one of three treatments: (1) injection of LPS, (2) Tail-docking or (3) Handling. When the ewes lambed (aged 2 years) behavioural (behaviour, ear-postures, vocalisations) and physiological (heart rate, heart rate variability, surface body temperature) parameters were measured during a 10-minute pre-treatment period and a 10-minute post-treatment period whilst the ewes were exposed sequentially to each of four test conditions): (Day 2) Control 1 (ewes left undisturbed with their lambs); (Day 3) Handling (H) (lambs weighed); (Day 3) C/TD (lambs castrated and/or tail-docked using a rubber ring) and (Day 6) Control 2 (ewes left undisturbed with their lambs). Data were analysed using a mixed between-within subjects ANOVA. H and C/TD had no effect on heart rate, heart rate variability and eye temperature of ewes. However, H and C/TD induced an increase in ear temperature, lamb-directed behaviour and number of ear-posture changes, and this response was significantly greater following C/TD than H. Low-pitched bleating and proportion of time spent with an asymmetrical ear-posture increased only in response to C/TD. The ewes' neonatal experience had no significant effect on their response to H and C/TD. Although the discrepancies between the behavioural and physiological responses of the ewes makes it uncertain whether ewe welfare is reduced by C/TD, the presence of behavioural changes indicates that ewes are responsive to their lambs' exposure to handling and a painful procedure.

Redirected behaviour and displacement activities in learning tasks: the commercial laying hen as model

Kuhne, Franziska, Adler, Silke and Sauerbrey, Anika Frauke Christine, Animal Welfare and Ethology, Department of Veterinary Clinical Sciences, Justus-Liebig University Giessen, Frankfurter Strasse 104, 35392 Giessen, Germany; franziska.kuhne@vetmed.uni-giessen.de

Redirected behaviour and displacement activities occur when some course of action is thwarted or inhibited (frustration). They also occur as adjunctive behaviours in operant conditioning tasks, where they might reflect frustration about unrewarded responses. Because frustration is associated with stress, and stress may interfere with learning and memory, we studied whether the occurrence of adjunctive behaviour is correlated with learning success in a series of visual-cue discrimination tasks. Twenty-one hens, aged 40 weeks were tested on acquisition, reversal, extinction, and relearning. The experimenters randomly assigned red and blue paperboard discs as discriminative stimuli. The learning criterion for each task was 90% correct responses in 20 trials in two consecutive test sessions. In the extinction session, the hens had to learn to inhibit pecking at a disc for a period of ten seconds, i.e. the maximum time for each extinction trial was set at ten seconds. The behavioural reactions were analysed using the repeated measures analysis of variance within the linear mixed models procedure of the SPSS 17® statistical software package. The hens became very restless, attempted to escape, showed displacement preening and redirected pecking specially in the acquisition and extinction tasks. Pecks at the surroundings ($F(3,9)=14.17$, $P<.001$) and the duration of the bouts of preening ($F(3,2)=11.43$, $P<.05$), but not the frequency of preening significantly correlated with learning success, in that the more the hens showed these adjunctive behaviours, the more trials they needed to meet the learning criteria. Thus, laying hens are susceptible to the effects of frustration elicited by operant procedures in visual discrimination tasks. The learning success in visual-cue discrimination tasks is correlated with the occurrence of adjunctive behaviour that can be deduced from the underlying motivation and suppression of previously learned responding.

Fear response in gestating sows

Phillips, Christina[1], Li, Yuzhi[1] and Anderson, Jon[2], [1]Univeristy of Minnesota, Animal Science, 335F AnSci/VetMed, 1988 Fitch Ave., St. Paul, MN 55108, USA, [2]University of Minnesota, Science and Mathematics, 1330 Sc, 600 E. 4th St., Morris, MN 56267, USA; christinaphillips@live.com

Fear, as an emotional state, can induce chronic stress, alter behaviors and reduce productivity. Both physiological status and experience can impact fear response of an animal. The objective of this study was to determine effects of gestation stage and age on the fear response of sows. Sixty gestating sows (parities 1 to 9) were subjected to the human approach and the novel object fear tests in early (2 wk after breeding) and late (12 wk after breeding) gestation periods. In the human approach test, the sows were individually moved to a test arena and given 2 min to be familiarized. A person then quietly entered the arena, and stood stationary at the end of the arena for 3 min, during which sow interactions with the person were recorded (time taken to approach within 0.5 m of the person, time spent within 0.5 m of the person, the number of interactions with the person, and time to the first touch with the person). Following the human approach test, a novel object was placed at the end of the arena. During the next 3 min, sow interactions with the novel object were recorded. The fear score for each sow was calculated using the Principle Component analysis, which suggested a single-dimension fear score. Fear scores ranged from 0 to 4.93, with lower scores being less fearful. The Glimmix procedure of SAS was used to test effects of parity and gestation period on fear response. Sows in late gestation were more fearful than sows in early gestation (3.17 vs. 2.39, SE = 0.30; $P<0.01$). There was no difference in fear score between young (parity 1-5) and old (parity 6-9) sows. These results indicate that gestation stage affected fear response in sows, which may be associated with changes in hormone profile during gestation.

Non-invasive assessment of positive emotions in horses using behavioural and physiological indicators

Stratton, Rachael[1], Waran, Natalie[2], Beausoleil, Ngaio[1], Stafford, Kevin[1], Worth, Gemma[3], Munn, Rachel[1] and Stewart, Mairi[3], [1]Massey University, IVABS, Palmerston North, New Zealand, [2]Unitec Institute of Technology, Department of Natural Sciences, Auckland, New Zealand, [3]AgResearch Limited, Animal Behaviour and Welfare Group, Hamilton, New Zealand; R.B.Stratton@massey.ac.nz

Recent research has investigated positive emotions (e.g. pleasure) in sheep and dogs in response to shifting animal welfare science focus towards optimising welfare through the presence of positive states. We investigated responses of thirteen horses during four putatively positive and negative treatments, using a Latin square design; P1 (grooming preferred site), P2 (clicker noise previously paired with feeding), N (neutral, no treatment) and NV (grooming non-preferred site). A 24-hour rest period was given between treamtents. Behaviour, heart rate (HR) and respiratory rate (RR) were recorded continuously pre-treatment (10min B), during (5min D) and post-treatment (1st 5min A1, 5-10min A2). Preliminary analysis was done using REML and ANOVA. For all treatments, there was no difference in behaviour change when the treatment period (D) or the immediately post-treatment period (A1) was compared to pre-treatment (B) because of high individual variability. However, in the 5-10 min after the end of treatment, change in both head and leg movements varied between treatments (A2 minus B, $P=0.056$, $P=0.044$ head and leg respectively), with P1 increasing movements while N, NV and P2 led to a decrease. There was a treatment effect on change in RR during treatment compared to pre-treatment (D minus B, $P=0.006$) with NV increasing RR (2.4 ± 0.63bpm). HR change during the first 12 seconds of treatment (relative to 60 seconds pre-treatment) varied according to treatment ($P=0.034$) with P1 and P2 displaying a spike. There was a treatment effect on HR change during 1-5 min of treatment ($P<0.001$), however, HR rose with NV and N but dropped with P1 and P2. Based on this preliminary analysis, post-treatment behaviour and HR druing treatment show potential for assessing putatively positive experiences in horses. In addition, RR seems primaril useful in differentiating the negative treatment.

Positive interactions lead to long term positive memories in horses (*Equus caballus*)
Sankey, Carol, Richard-Yris, Marie-Annick, Leroy, Hélène, Henry, Séverine and Hausberger, Martine, Ethos / UMR CNRS 6552 - Université de Rennes1, Station Biologique, 35380 Paimpont, France; carol.sankey@univ-rennes1.fr

Social relationships are important in social species. These relationships, based on repeated interactions, define each partner's expectations during the following encounters. The creation of a relationship implies high social cognitive abilities which require that each partner is able to associate the positive or negative content of an interaction with a specific partner and to recall this association. In this study, we tested the effects of repeated interactions in a training context on the memory kept by 23 young horses about humans, after 6 and 8 months of separation. They were divided in two groups: horses trained to remain immobile on a vocal command with positive reinforcement, which was a food reward (PR, N=11) and controls, trained by simple repetition of the task with no reinforcement, nor punishment (C, N=12). After learning the immobility command, horses from both groups underwent several handling procedures using the same command. Horses trained with the reward required less time to complete training than controls ($X_{PR}=3.75\pm0.08$ h, $X_C=5.24\pm0.24$, MW, $P<0.001$) and six months after the end of training, they remained immobile for longer than controls when given the vocal order ($X_{PR}=55.8\pm2.2$ s, $X_C=38.0\pm6.3$, MW, $P<0.05$). The association of a reward with a learning task in the training context induced positive reactions towards humans (e.g. sniffing the trainer: $X_{PR}=95.5\pm14.4$ occurrences, $X_C=57.5\pm.8.5$, MW, $P<0.05$). It also increased contact and interest, not only just after training, but also several months later (Time spent close to the trainer 6 months after training: $X_{PR}=59.8\pm4.6\%$, $X_C=15\pm5.6$, MW, $P<0.001$), despite no further interaction with humans. In addition, this 'positive memory' of humans extended to novel persons (Time spent close to an unknown person: $X_{PR}=39\pm6.4\%$, $X_C=7.2\pm3.2$, MW, $P<0.01$). Overall, these findings suggest remarkable social cognitive abilities that can be transposed from intraspecific to interspecific social contexts.

Grouping after parturition in presence of kids reduces stress in young dairy goats

Szabò, S.[1], Barth, K.[2], Graml, C.[1], Futschik, A.[3], Palme, R.[4] and Waiblinger, S.[1], [1]University of Veterinary Medicine Vienna, Institute of Animal Husbandry and Welfare, Veterinärplatz 1, 1210 Vienna, Austria, [2]Institute of Organic Farming, Johann Heinrich von Thünen Institute, Federal Research Institute for Rural Areas, Forestry and Fisheries, Trenthorst 32, 23847 Westerau, Germany, [3]University of Vienna, Department of Statistics, Universitätsstrasse 5/9, 1010 Vienna, Austria, [4]University of Veterinary Medicine Vienna, Institute of Biochemistry, Veterinärplatz 1, 1210 Vienn, Austria; Susanne.Waiblinger@vetmeduni.ac.at

Introduction of replacement goats into the adult dairy herd may cause considerable stress. The aim of this experiment was to compare two different times of grouping used in practice with regard to the level of stress experienced by the goats. A total of 32 young dairy goats where introduced in groups of 4 animals into one of two herds of 37 or 38 adult goats either during the dry period of the herd (young and adult goats were pregnant = DRY, n=16 young goats in 4 groups) or shortly after parturition (all animals lactating and with their kids = KIDS, n=16 young goats in 4 groups). Each group of young goats stayed in the adult herd for one week during which social interactions were recorded by focal continuous sampling and nearest neighbours of young goats by scan sampling every 10 min (220 scans per goat). Faeces of young goats was sampled for analysing cortisol metabolite concentrations at day -2,-1, 3, 5, 7 of introduction. Data were analysed by LMM. Young goats received more agonistic behaviours with physical contact during DRY than KIDS ($P=0.000$, $F=17.5$). Young goats were each others nearest neighbours far above chance levels in both periods but even more distinct during DRY (mean±SE 79±0.008% of scans, KIDS: 55±0.009%, $P=0.000$). Cortisol metabolite levels were increasing during the introduction in DRY reaching their peak on day 3 ($P=0.000$), while hardly any changes were seen during KIDS (difference DRY- KIDS: $P=0.000$, $F=56.5$). The results suggest that introducing young goats into a herd of adult dairy goats shortly after parturition with kids still present induces less stress and thus is preferable. The effect of kids vs. stage of pregnancy needs further investigation. We acknowledge funding by the BMLFUW and BMGF, Project Nr. 100191.

Undernutrition during pregnancy impairs the maternal response to the young during a double choice recognition test in 12 old hours goat kids

Pelayo, Brenda, Soto, Rosalba, Gónzalez, Francisco, Serafin, Norma, Vázquez, Luis and Terrazas, Angelica, Universidad Nacional Autonoma De Mexico, Facultad de Estudios Superiores Cuautitlan e Instituto de Neurobiologia, KM 2.5 Cuautitlan - Teoloyucan San Sebastian Xhala, 54714 Cuautitlan Izcalli, Mexico; garciate@servidor.unam.mx

In Mexico the goat´s production suffers of several periods with underfeeding that affects their reproductive stages. The aim of this experiment was to study the effect of undernutrition during pregnancy on maternal behavior response to the young during a double choice recognition test carried out to 12 hr kids. French-Alpino goats were used in two groups: Control (n=7), fed with 100% of their nutritional requirements (C) and Underfed (n=8) fed with 70% of their requirements from day 75 of pregnancy until the birth (U). Twelve C and 11 U kids were tested in a 5 minutes double choice test between alien and own dams. Control dams did more attempts of jumping out than U goats (2.1 ± 0.4 vs. 0.6 ± 0.2, $P=0.005$). The own C mothers had more frequency to with their head down than up (4.2 ± 0.6 vs. 0.83 ± 0.38, $P=0.002$). Aliens C mothers tended to showed more frequency to be head up than own mothers ($P=0.08$).Own C mothers had more frequency to be with their head down than alien (4.2 ± 0.6 vs. 2.1 ± 0.4, $P=0.05$). Non difference of this comparisons were found in U dams ($P>0.05$). In both groups the dams emit more low pitch-bleats than high-pitch bleats ($P<0.01$). Non differences were found between C and U kids. However C kids spent longer time looking toward the own dam than to the alien (52.6 ± 10.3 vs. 22.6 ± 6.5 sc. $P=0.04$) and tended to spend more time near the own than alien dam (76.2 ± 25 vs. 32 ± 16.7 sc, $P=0.1$). While U kids did not show preferences for any dam ($P>0.05$). Maternal underfeeding during pregnancy affects the mother response to the kid in a double choice test that could impair the normal mother-young bonding. Supported by UNAM-PAPIIT IN207508.

Behavioural indicators of analgesic efficacy in rabbits under two different analgesic regimens

Farnworth, Mark J.[1], Schweizer, Katharine A.[2], Walker, Jessica K.[1], Guild, Sarah-Jane[3], Barrett, Carolyn J.[3] and Waran, Natalie K.[1], [1]Unitec Institute of Tech., Department of Natural Sciences, Building 115, Carrington Rd, Mt. Albert, Auckland, New Zealand, [2]University of Edinburgh, Royal (Dick) School of Veterinary Studies, Easter Bush Vet. Centre, Easter Bush, Roslin, EH25 9RG, United Kingdom, [3]University of Auckland, Department of Physiology, Pvt. Bag 92019, Auckland, New Zealand; mfarnworth@unitec.ac.nz

Behavioural information, using pre-established pain behaviours, was collected on twelve male New Zealand White rabbits to assess differences resulting from two analgesic protocols following abdominal surgery. Subjects were sequentially divided into two treatment groups. Group 1 (PO n=6) received standard peri-operative analgesia, 2mg/kg of the non-steroidal anti-inflammatory drug Carprofen 30 minutes prior to surgery. Group 2 (PO+ n=6) received additional post-operative Carprofen (2mg/kg) prior to recovery from anaesthesia. In accordance with best practice all individuals received carprofen (2mg/kg) 24h post-surgery. Observations were separated into four time periods: T1 immediately post-surgery (0-4hrs); T2 (+4-8hrs) T3 (+21-24hrs) and T4 (+24-27hrs). Behavioural analysis was performed using Observer XT. Kruskal-Wallis tests determined differences between the treatment groups. Friedman's tests assessed between time periods. Results showed no clear significant differences in expression of pain behaviours between PO+ and PO groups in any time period. However during the first 27 hours post-operation all individuals expressed elevated levels of established pain behaviours (e.g. 'tight huddle' X^2=12.7; df=3; $P<0.005$) whilst normal maintenance behaviours, such as eating (X^2=22.842; df=3; $P<0.001$) and grooming (X^2=12.3; df=3; $P<0.006$), remained suppressed. The group concludes that the additional post-operative analgesic protocol used here appears not to provide additional pain relief beyond that provided by the standard peri-operative protocol. However behavioural changes suggest that some pain continues to be experienced. Research into multimodal analgesia to assess and manage the potential for non-inflammatory pain (visceral and somatic) following abdominal surgery may prove beneficial for pain management, and subsequent welfare, of rabbits.

'Home-alone' dogs sleep more when played classical music
Rutter, S. Mark and Johnson, Fiona, Harper Adams University College, Animals Department, Newport, Shropshire, TF10 8NB, United Kingdom; smrutter@harper-adams.ac.uk

Many companion dogs are often left alone at home, and this separation from their owner can lead to stress and behavioural problems. Previous studies with dogs in rehoming centres have shown that classical music had a calming effect on their behaviour. This study investigated the effects of classical music on 'home-alone' dogs. Four dogs each received four treatments (in a Latin Square) over four consecutive days for four hours a day in their home environment whilst their owners were out. The treatments were: talk radio (TR), classical music radio (CR), classical music compilation (CC) and silence (S). The radio treatments were pre-recorded so that each dog received identical radio programmes. The CC treatment was a pre-recorded sequence of 'relaxing' classical music. Two video cameras were used to record the dog's behaviour. These recordings were analysed using instantaneous sampling at five minute intervals and the following behaviours were recorded: posture (lie, sit, stand, walking, running), vocalisations, behavioural activity (sleeping, resting awake, play, groom, social behaviour, destructive behaviour). Sleeping was defined as lying down with eyes closed and resting awake was lying with eyes open. ANOVA showed a significant effect of treatment on sleeping ($F_{3,45}=5.03$, $P=0.004$) and resting awake ($F_{3,45}=4.99$, $P=0.005$). Dogs spent a statistically significantly greater percentage of the 4hr period sleeping during the CR (78%) and CC (82%) compared with S (60%), and less time resting awake during CR (21%) and CC (18%) compared with S (40%) (post-hoc Tukey's test). These results showed that dogs slept (rather than resting awake) more when classical music was played compared with silence. Although requiring further research, sleeping could be interpreted as more beneficial than resting awake, therefore the calming effect of playing classical music could be favourable to the welfare of dogs whilst separated from their owner.

Animal welfare education in Austria: assessing student and teacher responses to 'Tierprofi-Nutztiere'

Liszewski, Melissa[1], Heleski, Camie[1], Prevost, Bärbel[2], Scheib, Marie-Helene[3] and Winckler, Christoph[2], [1]Michigan State University, 1290 Anthony Hall, East Lansing, MI 48824, USA, [2]University of Natural Resources and Applied Life Sciences, Gregor-Mendel-Strasse 33, 1180 Wien, Austria, [3]Verein 'Tierschutz macht Schule', Maxingstraße 13b, 1130 Wien, Austria; mel.liszewski@gmail.com

Austria's federally funded 'Tierschutz macht Schule' program has provided animal welfare educational materials to >900 schools nationwide. Much time/effort have gone into the development/dissemination of materials but until now it has been widely unknown if these resources are actually being utilized after distribution and if students and teachers find them at all valuable/interesting. To address these unknowns we conducted a pilot study assessing the outcomes of classrooms who had received the Tierprofi-Nutztiere farm animal welfare booklet, conducted via written teacher/student questionnaires distributed to 8 schools in Vienna. Questionnaire responses gathered from 108 students (ages 8-13) and 14 teachers were analyzed primarily using percentages for quantitative data and content analysis for qualitative data. Responses indicated that students and teachers are utilizing Tierprofi-Nutztiere and regard knowledge of farm animals and their welfare as interesting/valuable: 82% of students were interested in learning about farm animal welfare while 94% of students and 100% of teachers found this type of education important; 88% confirmed they had learned new information from the booklet and 72% wanted to learn more. Further results suggest however that many students still cannot fully grasp the deeper relationship between consumer behavior and animal welfare and fully connect products they consume to actual animal beings: 86% prefer eating meat from chickens pictured in a free range setting but only 56% could correctly trace all animal products on a pizza back to their animal origins; while 58% thought their choice of products to purchase/consume could improve the lives of farm animals only 17% believed consumers decide how farm animals live. Results confirm that a full scale study is warranted in order to test an extended hypothesis that students and teachers utilizing Tierprofi-Nutztiere are actually improving their knowledge of/attitudes towards farm animals and their welfare.

The relationship between empathy, perception of pain and attitudes toward pets among Norwegian dog owners

Ellingsen, Kristian[1], Zanella, Adroaldo Jose[2], Bjerkås, Ellen[2] and Indrebø, Astrid[2], [1]National Veterinary Institute, Animal Health and Welfare, Pb 750 Sentrum, 0106 Oslo, Norway, [2]Norwegian School of Veterinary Science, Dept. Production Animal Clinical Sciences, P.O. Box 8146 Dep., 0033 Oslo, Norway; kristian.ellingsen@veinst.no

Anthropomorphism, attachment level, belief in animal mind, the owner's level of empathy, and attitudes toward pets are important factors affecting human-animal interactions. In addition to characterizing Norwegian dog owners, the aim of this work was to study the relationship between empathy, attitudes and pain. Study population consisted of 1896 Norwegian dog owners sampled using an Internet based survey (QuestBack ™). The questionnaire included the Pet Attitude Scale (PAS), the Animal Empathy Scale (AES) and demographic questions. Participants were also presented with 17 photos, provided by the Norwegian School of Veterinary Science, showing dogs experiencing painful conditions and asked to rate the level of pain of which they believed that animal was enduring, using Visual Analogue Scales (VAS). Data were analysed using independent samples t-tests, Pearson's linear correlation, partial correlation, and Principal Component Analysis. Women scored higher than men (mean & SD) for PAS (116±6.6 × 119.4 ± 9.9) for AES (119.5±12.9 × 103.1±14.3) for VAS pain pictures (12.5 ±2.6 × 11.5±2.7) ($P<0.001$). Participants who reported childhood pet keeping scored higher than those without childhood pet experience for PAS (114.4±7.9 × 112.1 ±11.1) and AES (114.5 ±15.1 x 109.7±17.4) ($P<0.001$). Income and education were negatively correlated with all three instruments ($P<0.01$). People who reported keeping their dog primarily for companionship scored higher on the PAS, AES and pain scenarios ($P<0.001$) compared to gundog owners. A positive correlation ($r=0.58$) was found between animal-directed empathy and positive attitudes toward pets. Empathy was found to be the best predictor of how people rated pain in dogs, however the moderate correlation ($r=0.31$) indicates that other processes are involved. This study offered valuable information on the relationship between empathy, attitudes and pain, which will facilitate the development of welfare training programs for dog owners.

Barrier perches and density effects in broilers

Ventura, Beth[1], Estevez, Inma[1,2] and Siewerdt, Frank[1], [1]University of Maryland, Animal and Avian Sciences, Regents Drive, College Park 20742, Maryland, USA, [2]IKERBASQUE & NEIKER-TECNALIA, Animal Production, P.O. Box 46, Vitoria-Gasteiz, 01080, Spain; bethventura2@gmail.com

Restriction of opportunities to express natural behavior is common in commercial broiler production, especially when birds are reared at high densities. We hypothesized that this welfare concern can be addressed by the provision of low barrier perches, which can encourage perching behavior and increase environmental complexity across a range of densities. In this experiment, 2,088 day-old broiler chicks were randomly assigned to one of the following barrier and density treatment combinations over four replications: simple barrier, complex barrier, or control (no barrier) and low (8 birds/m2), moderate (13 birds/m2), or high (18 birds/m2) density. Behavioral data were collected on focal birds via instantaneous scan sampling from 2 to 6 weeks of age. Mean estimates per pen for the percent of observations performing each behavior were quantified and analyzed with a mixed model ANOVA with week as the repeated measure. Our results showed that the behavioral time budget was affected by barrier perches, density, and age. Both barrier perches effectively stimulated high perching rates, and also reduced aggression and disturbances relative to controls ($P<0.05$). Compared to control treatment, birds fed less often in complex barrier pens ($P<0.001$) and drank less often in simple barrier pens ($P<0.01$), though final body weights, feed conversion and mortalities were unaffected by treatment. Density had a suppressive effect on activity, with lower perching ($P<0.0001$) and more frequent sitting ($P=0.001$) occurring earlier in the rearing period at 18 birds/m2 compared to the lower densities. High densities were also accompanied by a decline in walking ($P<0.05$) and foraging ($P<0.005$) and an increase in disturbances ($P<0.05$). Our results demonstrated that increasing density compromises bird welfare by promoting inactivity and increasing disturbances. We suggest that barrier perches have the potential to improve broiler welfare by decreasing aggression and disturbances and by encouraging activity, notably by providing accessible opportunities to perch.

Welfare implications of methods used to kill wildlife

Cockram, Michael S., Atlantic Veterinary College, University of Prince Edward Island, Sir James Dunn Animal Welfare Centre, Department of Health Management, 550 University Avenue, Charlottetown, PEI, C1A 4P3, Canada; mcockram@upei.ca

Wild animals are killed for a variety of reasons including subsistence, commercial and sport hunting, disease control, management culling and pest control. The methods used vary considerably and the potential welfare implications associated with each method vary according to the manner of death, the species of animals and the skill of the people using them. The risk of suffering (e.g. fear, distress and pain) before death, the time to loss of consciousness, the effectiveness of killing and the risk of escape of wounded or injured animals need to be considered. Studies that provide evidence on the stress associated with killing, the manner in which death occurs and the rate of wounding will be reviewed. Using shooting with firearms as an example, the occurrence of stress before death, the effects of target location and type of ammunition on the effectiveness of killing and the rate of wounding will be discussed. Published work on the use of firearms and/or a blow to the head to kill seals will be evaluated. Studies that have recorded the behaviour, stress and injury in animals either killed or restrained in traps before death will also be reviewed. This topic has not been studied in the same detail as the slaughter of domesticated animals for food and the existing literature is scarce and in diverse sources. However, some methods of killing wildlife have caused public concern and the suffering associated with some methods may be considerable and affect large numbers of animals. In situations where there are differences of opinion on the treatment of animals and a need for policy development, a review of peer-reviewed scientific evidence is beneficial.

Onset of the breeding season in goats is advanced by the presence of high-libido bucks

Delgadillo, José Alberto, Universidad Autónoma Agraria Antonio Narro, Ciencias Médico Veterinarias, Periférico Raúl López Sánchez y Carretera a Santa Fe, 27054 Torreón, Coahuila, Mexico; joaldesa@yahoo.com

The objective of this study was to determine the importance of male sexual behavior in the onset of the breeding season in goats. Two male goats were exposed to natural photoperiod variations; other two males were exposed to 2.5 months of long days from November 1 to stimulate their sexual activity during the non-breeding season. One group of does (n=7) remained isolated from males. On February 1, two other groups of does (N=7 each) were exposed to light-treated and light-untreated males. Courtship behaviors from each male were recorded in three consecutive days during one hour h in March (non-breeding season) and May (breeding season). Presence of corpus luteum were determined twice weekly by ultrasound. Frequencies of sexual behaviors were compared using the Fisher exact test and an ANOVA. The onset and end of ovulatory activity were compared using a two-way ANOVA. In March, the light-treated bucks displayed higher frequencies of ano-genital sniffing (250 vs. 3), nudging (350 vs. 15), and mount attempts (10 vs. 0) than untreated ones ($P<0.05$). More ano-genital sniffing (160 vs. 3), nudging (280 vs. 15) and mount attempts (13 vs. 0; $P>0.05$) were displayed by the untreated males in May than in March ($P<0.05$). The end of the breeding season did not differ between groups, and occurred between 15 and 25 February ($P>0.05$). In contrast, the first ovulation occurred earlier in females exposed to the light-treated males (March 23 + 3 days) than in those exposed to untreated males (June 26 + 6 days) or isolated does (September 20 + 8 days; $P<0.05$). Male sexual behavior is more important than the presence of males to advance the onset of the breeding season in female goats.

Dog tail-chasing behaviour and human responses to it: insights from YouTube

Burn, Charlotte C and Browning, Verity J, The Royal Veterinary College, Veterinary Clinical Sciences, Hawkshead Lane, Al9 7TA, United Kingdom; cburn@rvc.ac.uk

Tail-chasing in dogs can indicate veterinary problems, including canine compulsive disorders, or tail or hind-quarter discomfort. Few data exist beyond case studies, because tail-chasing is intermittent and only reliably performed by severely affected dogs. We aimed to gain information on its behavioural characteristics, context, and human awareness of potential welfare implications. We recorded behaviour characteristics, context, tail morphology, and YouTube user comments from the first 400 of 3340 video hits for 'dog chasing tail', plus breed-matched non-tail-chasing control videos. YouTube users form a non-random, voluntary sample population, but their comments benefit from being unprompted by our study purpose. Hypotheses were tested using generalised models with ID, breed, and breed group as random factors. Dogs comprised 68 breeds, and spinning speed could be rapid: mean (s.e.) = 42.6 (1.2) spins/min. Tail-chasing included tail mouthing/biting in 63% of videos, and object collisions in 26%, and 71% of dogs appeared difficult to distract during bouts. Tail-chasing incorporated play behaviour in 17%, especially in younger dogs (t = 4.89; DF = 105; $P<0.001$). Tail-chasing videos were less frequently outdoors than controls (t = 13.01; DF = 399; $P<0.001$). Dogs in tail-chasing videos had longer tails (t = 2.15; DF = 196; $P= 0.033$), less frequently docked (t = 10.66; DF = 396; $P<0.001$), than controls. Laughter was heard in 55% of videos, and 59% were described as 'funny', 26% 'crazy', and 15% 'stupid'; 2% had comments suggesting veterinary explanations. Verbal and/or physical encouragement was observed in 43% of videos. As with most surveys, the generalisability of the conclusions is limited by study population biases, as will be discussed. Nevertheless, results suggest that while tail-chasing can comprise play, it can frequently become perseverative; YouTube users showed little awareness of animal welfare implications, and frequently demonstrated inappropriate encouragement.

Farmer's attitude affects piglet production parameters

Kauppinen, Tiina[1], Valros, Anna[2] and Vesala, Kari Mikko[3], [1]University of Helsinki, Ruralia Institute, Lönnrotinkatu 7, 50100 Mikkeli, Finland, [2]University of Helsinki, Faculty of Veterinary Medicine, P.O. Box 57, 00014 University of Helsinki, Finland, [3]University of Helsinki, Faculty of Social Sciences, P.O. Box 54, 00014 University of Helsinki, Finland; tiina.kauppinen@helsinki.fi

Production animal welfare and even productivity are greatly influenced by the quality of stockmanship. Farmers' attitudes affect their behavior and predict their actions towards animals. Our aim was to find welfare-related attitudes that also correlate with animal productivity. We studied farmers' attitudes towards animal welfare through a survey with 300 (37%) respondents. In this data (analysed with PCA), animal welfare was organized at two conceptual levels. At a concrete level, the farmers perceived it as treating the animals humanely, providing the animals with a favorable environment, enhancing the farmer's own well-being, and taking care of the animals' health. At an abstract level, the respondents were profiled as reward-seeking or as empathic farmers. They also mentioned several subjective norms that they perceived important. Correlations between the pig farmers' attitudes (N = 124) and standardized piglet production parameters showed that the farmers emphasizing the importance of the humane treatment weaned more piglets compared with a Finnish average (partial correlation, $P= 0{,}25^{*}$). Farmers who perceived it easy to provide the animals with a favorable environment also weaned more piglets ($P= 0{,}26^{**}$) and had a lower piglet mortality before weaning ($P= 0{,}27^{*}$). Perceiving slaughterhouses, agricultural advisers and researchers as important norms was associated with a higher number of stillborn piglets ($P= 0{,}22^{*}$, $P= 0{,}20^{*}$, $P= 0{,}20^{*}$). Yet the perceived importance of researchers correlated with lower piglet mortality ($P= 0{,}22^{*}$), larger litters ($P= 0{,}26^{*}$) and higher number of weaned piglets ($P= 0{,}27^{*}$). Our results show that farmers' attitudes count: treating the animals humanely, investing in a favorable environment, and having a positive attitude towards new information and scientific research is associated with an above-average productivity on piglet farms. These attitudes, when implemented and concretized in practice, also benefit the animals through a higher standard of welfare.

The effect of age of young calves on the ease of movement through an obstacle course

Jongman, Ellen, Animal Welfare Science Centre, Department of Primary Industries Victoria, 600 Sneydes Road, Werribee 3030, Australia; ellen.jongman@dpi.vic.gov.au

Young calves are vulnerable to the stressors of handling and transport, and can be particularly difficult to move after transport. This study examined ease of handling of calves at 3, 5 and 10 days of age. Calves were moved through a 23.4m unfamiliar 'obstacle course' consisting of a ramp of 12° incline (Zone 1), along a platform (Zone 2), 2 turns of 90° (Zone 3) and a ramp of 11° decline (Zone 4). Over a 5-week period a total of 49 calves were randomly assigned and tested once at 3, 5 or 10 days of age. The calves were moved by the same handler, who walked behind the calf, and intervened whenever the calf was stationary for more than 3s. Intervention consisted of a gentle hold or push, using one or two hands at the rear or the flank (A). If the calf did not walk voluntarily after being encouraged in this manner for 3s, a push at the rear strong enough to move the calf was used (P). Interactions between the handler and the calf and the time taken to move through each section were recorded. The average time in each zone of all calves tested on the same day in the same calf age grouping, was calculated. After square root transformation, these averages were analysed using REML analysis. There was a significant effect of age on ease of handling. Older calves moved through the obstacle course faster than younger calves (99, 86 and 72s for 3, 5 and 10 day old calves, respectively, $P<0.05$). Older animals moved especially faster through Zone 3 ($P<0.001$) and Zone 4 ($P<0.05$) and needed less assistance (A) throughout the obstacle course ($P<0.05$). The age of young calves may affect ease of handling, which may have implications for handling associated with transport and slaughter.

On farm assessment of general fearfulness in growing pigs

Dalmau, Antoni and Velarde, Antonio, IRTA, Animal Welfare Unit, Finca Camps i Armet S/N, 17121 (girona), Spain; antoni.dalmau@irta.es

The aim of the project was to test different measures to assess fearfulness in growing pigs on commercial farms. A novel object test was carried out in 321 pens of 18 farms. The novel object consisted of three balloons (one red, blue and yellow) filled with helium and tied with a weight to maintain them at the pig's eyes position when standing. The behaviour of the animals was assessed according to: (1) The time of the first pig to touch any of the three balloons, and, (2) the percentage of pigs into the pen that were looking directly at the balloons. Statistical analyses were carried out with proc Genmod of SAS. Significance was fixed at $P<0.05$. The time taken to touch the balloons was different between farms and ranged from 6.8±0.66s to 73.3±24.35s. When differences were studied between different buildings in the same farm, there were found differences only in one farm. In addition, there were found differences among pens located in a same building only in two cases (5% of the total buildings studied). The percentage of animals watching directly to the balloons in the 321 pens studied was 24.0± 0.20%. The farm with the lowest percentage showed a value of 12.7 ± 0.47% and the farm with the highest a value of 32.2 ± 0.82%. Most of the farms with different buildings showed differences between buildings. In addition, in 33 of 39 buildings there were found differences between pens located in the same building. The time taken for touching the balloon is more robust (few differences between pens into a same farm) than the animals watching the balloons. In consequence, to study the fearfulness of growing pigs on farm the time taken for touching the balloons is recommended.

Improving grazing livestock welfare at marking in Australia

Windsor, Peter, Lomax, Sabrina, Espinoza, Crystal and Sheil, Meredith, University of Sydney, Faculty of Veterinary Science, Werombi Road, Camden, NSW 2570, Australia; peter.windsor@sydney.edu.au

Managing extensively raised livestock in Australia requires routine lamb and calf marking procedures that are laborious for producers and painful for livestock. This includes vaccination, ear tagging/knotching, castration of most males of both species, tail-docking of all sheep and mulesing of many wool sheep (removal of perineal skin to decrease the risk of blowfly strike) and dehorning of many cattle breeds. Marking of grazing livestock almost never directly involves veterinarians in Australia or includes pain relief, but is justified by graziers as it enhances animal management. Consumer concerns with mulesing adversely affected the international marketing of Australian wool, forcing industry to promote cessation of mulesing by 2010. This is now considered unrealistic, with the preferred long term solution being selection of sheep that are less susceptible to flystrike, a process that may take decades. In the interim, we have shown that lambs treated at mulesing with farmer applied spray-on topical anaesthetic formulation (Trisolfen®, Bayer Animal Health, Australia) demonstrated significantly lower pain-related behavior scores compared with placebo gel treated ($P=0.03$), and untreated mulesed lambs ($P<0.001$) and were not significantly different from unmulesed controls. Further, we demonstrated that lambs treated with topical anaesthesia at castration and tail-docking displayed significantly less pain-related behaviors than untreated, ring-castrated and placebo-treated lambs ($P<0.001$). We then attempted to address concerns that animals should have analgesia prior to marking, by low dose (50µg/kg) aplication of the sedative xylazine and/or therapeutic levels of the non-steriodal anti-inflammatory agent carprofen applied prior to surgeryt, with or without spray-on topical anaesthesia. Results of these experiments to further enhance welfare, plus discussion of modifications of our techniques for assessing pain in young livestock, will be presented.

Stress assessment in captive white-lipped peccaries (*Mammalia, Tayassuidae*)

Nogueira-Filho, Sérgio L.G.[1], Carvalho, Heneile[1], Silva, Hélderes P.A.[2], Fernandes, Luis C.[2], Carvalho, Marco A.G.[1] and Nogueira, Selene S.C.[1], [1]Universidade Estadual de Santa Cruz, Laboratório de Etologia Aplicada, Rod. Ilheus Itabuna, km 16, 45662-900, Brazil, [2]Universidade Federal do Rio Grande do Norte, Laboratório de Endocrinologia Comportamental, Departamento de Fisiologia, Centro de Biociências, Caixa Postal 1511, 59078-970, Brazil; sergio.luiz@pq.cnpq.br

Records of peccaries (*Tayassu pecari*) that have reproduced in captivity are few and may be due to distress. The glucocorticoid metabolite concentrations may reflect animals' stress. These metabolites are present in feces and fecal glucocorticoid assays have been employed as a noninvasive technique to analyze animal distress. We tested this possibility by using an adrenocorticotropic hormone (ACTH) challenge in a fecal glucocorticoid assay. We tested the effects of three treatments: 0.25 or 0.50 mg.100kg^{-1} ACTH, and saline solution as a control, in six captive adult male white-lipped peccaries in a Latin Square design. We immobilized the animals and inject (I.M.) the treatments at 9:00AM of the challenge days. Feces were collected beginning 72 h prior to and ending 96 h after the injections. During the same period, we also observed each individual for 15-min with the continuous-recording based on focal animal sampling sessions. Feces were freeze-dried, extracted by an ethanol vortex method, and assayed for glucocorticoids using enzyme immunoassay. We compared the behavioral measures and the fecal glucocorticoids among the treatments by repeated measures ANOVA followed by Tukey tests. The peccaries remained more time on behavioral acts indicators of stress on the challenge days ($F_{4, 60}=14.72$, $P<0.0001$). The fecal glucocorticoid concentration in ACTH and saline control treatments did not differ among the treatments ($F_{2, 15} = 0.42$, $P= 0.67$) reaching a peak (265% above baseline) 24 h post challenge ($F_{4, 60} = 5.48$, $P= 0.008$) followed by a decline. Therefore, just the immobilization itself was meaningful stressor and ifluenced peccary's adrecortical rsponse to ACTH and saline control treatments. There are no control data from free range, thus we believe that the method of using the adrenocortical metabolites records of each animal as its own control, together with behavioral observations, will provide objective measures to monitor stress in white-lipped peccary.

Do laying hens learn to avoid feathers from conspecifics?

Harlander Matauschek, Alexandra[1], Beck, Philipp[1] and Piepho, Hans-Peter[2], [1]University of Hohenheim, Department of Small Animal Ethology and Poultry Science, Garbenstr. 17-470c, 70599 Stuttgart, Germany, [2]University of Hohenheim, Bioinformatics Unit, Institute for Crop Production and Grassland Research, Fruwirthstr. 23, 70599 Stuttgart, Germany; aharlander@gmx.at

Severe feather pecking is positively associated with feather eating, indicating that feathers are seen as a feeding substrate. There is evidence that birds can learn to avoid feed by adding a bitter taste (quinine). In the present experiment was investigated if laying hens avoid feathers when quinine is the aversive agent on the feather cover. At 21 weeks of age, laying hens were randomly divided into 12 groups of 10 birds each (6 groups of high feather-pecking (H) birds and 6 groups of low feather-pecking (L) birds) and kept in identical deep litter pens. In week 1 of the experiment all occurrences of severe feather pecking in a group were recorded for 20 minutes over a period of 4 consecutive days. In week 2 half of the groups feather cover was treated by spraying 4% quinine sulfate solution (Q) and the other half was kept as a control. Behavioural observations were performed in weeks 2, 3 and 4 as described in week 1. Data were analysed using mixed models. Differences for lsmeans estimates were tested using t-test. A significant decrease in severe feather pecking was detected in LQ birds in weeks 2 (lsmean 1.1, $P<0.0001$), 3 (lsmean 5.7, $P<0.004$) and 4 (lsmean 11.8, $P<0.001$) compared to week 1 (lsmeans 14.1). Likewise, in HQ birds, where a significant decrease of severe feather pecking was shown in weeks 2 (lsmean 48.5, $P<0.002$) and 3 (lsmean 27.1, $P<0.001$) compared to week 1 (lsmean 87.1). However, the quinine treatment showed a tendency of lowering severe feather pecking in week 4 (lsmean 39.1, $P<0.1$). Our experiment showed that birds were able to learn that feathers from conspecifics are not acceptable pecking substrate and to avoid them for a period of time.

Influence of fibre in diet on mating behaviour and reproductive success in female farmed mink

Spangberg, Agnethe and Malmkvist, Jens, Aarhus University, Animal Health and Bioscience, Blichers allé 20, 8830 Tjele, Denmark; a_spangberg@hotmail.com

Mink chosen for breeding are slimmed during winter to be ready for mating in the early spring. The slimming is induced by restricted feeding which increases the occurrence of stereotypic behaviour and may have a negative impact on welfare. Therefore, a fibrous diet is suggested used to increase the mink's satiety. This study aims to describe mating behaviour and the effects of two feed types on mating/reproductive success in farmed mink: (1) CONTROL: normal farm feed, (2) FIBRE: normal feed added fibre (barley peel), reducing the energy density with 26%. We focused on mating behaviour and reproductive success, using 336 first-parity mink females (CONTROL: N=169; FIBRE: N=167) of the same breeding line. The fibrous diet successfully reduced stereotypic behaviour during the winter period. The pre-mating concentration of cortisol metabolites (measured non-invasively in faeces) did not differ between treatments (CONTROL: 119 (11.8) ng/g, FIBRE: 106 (7.9) ng/g, $F_{1,299}=0.21$, $P=0.65$), and did not affect mating behaviour. Several aspects of mating behaviour depended on e.g. the female's previous mating success during the mating season. Male neck biting is a fundamental part of the mating repertoire, occurring in all successful matings. Overt aggression occurred at a low level (<0.5% of scanning observations) even in unsuccessful mating trials. Females that were mated twice delivered significantly more kits than females mated once ($F_{1,301}=7.0$, $P=0.009$). Heavier females at mating delivered more stillborns ($F_{1,302}=8.5$, $P=0.004$), and weight loss during the winter positively increased the litter size ($F_{1,301}=5.3$, $P=0.022$). However, large litters also had more stillborns ($F_{1,302}=33.9$, $P<0.001$), and contained smaller kits ($F_{1,301}=22.1$, $P<0.001$). In conclusion, we found no significant effects on mating behaviour and reproductive success after using the FIBRE feed.

Effects of breed, nutrition and litter size on fetal behaviour in sheep

Coombs, Tamsin M., McIlvaney, Kirstin M. and Dwyer, Cathy M., SAC, Animal Behaviour and Welfare Group, West Mains Road, Edinburgh EH9 3JG, United Kingdom; tamsin.coombs@sac.ac.uk

The function of fetal behaviour may be to establish muscular competence and move the fetus into the characteristic birth posture. In this study we investigated whether nutritional effects, litter size and breed differences seen in neonatal lamb behaviour are also seen in the fetus of two British sheep breeds, Suffolk (S) and Scottish Blackface (BF), using transabdominal ultrasonography. In experiment 1 control single-bearing ewes (C: n=12, BF=6, S=6) were fed 100% requirements for maintenance and fetal growth throughout pregnancy while restricted ewes (R: n=12, BF=6, S=6) were fed 75% of requirements from days 1 to 90 of pregnancy and 100% thereafter. In experiment 2 single- (Single: n=12, BF=6, S=6) and twin-bearing ewes (Twin: n=12, BF=6, S=6) were scanned. Ewes were scanned during 3 separate weeks (approximately gestational days 56, 77 and 98) for an average of 44 minutes per week. Fetuses were assigned a behavioural state score every 5 minutes during scanning as follows, (1) low activity: brief movements, mostly startles; (2) active: frequent movements, stretches and movement of head and limbs; (3) very active: Vigorous continual activity including trunk rotations. At d56 and d77 BF fetuses were significantly more active than S fetuses (Chi-square, d.f. =1, $P<0.05$) with more fetuses gaining scores of >1. At d56 SR fetuses tended to be more active than SC fetuses (Fisher's Exact, $P=0.06$) and at d77 S Single fetuses were significantly more active than S Twin fetuses ($P=0.04$). At d98 S fetuses changed between behavioural states significantly more often than BF fetuses (Mann-Whitney, S median =2, BF median =1, $P<0.05$). These data suggest that breed, nutrition and litter size, which affect neonatal lamb behaviour, also influence fetal lamb behaviour from early in gestation. Current research is investigating whether fetal behaviour is continuous with neonatal behaviour and to assess the implications for lamb survival.

Evaluation of the impact of otitis media on the behavior of dairy heifer calves

Stanton, AL[1], Leslie, KE[1], Widowski, TM[1], Leblanc, SJ[1], Kelton, DF[1] and Millman, ST[2], [1]University of Guelph, 50 Stone Rd E, Guelph, Ontario N1G 2W1, Canada, [2]Iowa State University, 1600 South 16th St., Ames, Iowa, 50011, USA; astanton@uoguelph.ca

The objective for this study was to identify the behavioural changes associated with disease in milk-fed dairy calves. Of particular interest was otitis media (ear droop), a prevalent infection of young calves in large calf rearing operations. Signs of otitis are unilateral (UNI) or bilateral (BIL) ear droop, which may progress to neurological signs. A total of 25 calves, at a heifer raising facility in New York State, were enrolled on this study twice monthly, following their arrival from 5 source farms. Calves were individually housed in naturally ventilated nursery barns. An accelerometer (Icetag®) was attached to the mid-metatarsal region of one hind limb to measure lying and step activity, and a smaller device (Actical®) was attached over the poll region to measure headshaking. Data was analyzed using generalized linear mixed models with repeated measures in SAS 9.1 Both devices recorded data for 28 days, and datasets were complete for 22 calves. UNI and BIL were identified by farm staff in 2 and 14 calves, respectively. Lying bouts, average lying duration, and total lying time were not significantly different in calves with ear droops compared to normal calves ($P=0.15$, $P=0.14$ and $P=0.73$, respectively). Calves with UNI performed 228 ± 102.6 fewer steps/day than normal calves ($P=0.02$). Actical® data was compared with steps to determine if there was an increased incidence of head shaking with otitis. These two measures were positively correlated ($R=0.64$; $P<0.001$). The Activity units/step did not differ between BIL and normal calves ($P=0.63$). These findings are consistent with the larger field trial, which found that UNI decreased average daily gain, but BIL did not. This may indicate differences in severity, symptoms, or rates of early detection. In conclusion, bilateral ear droop does not appear to have substantial affect on the behaviour of calves. Further research is needed to determine the impact of unilateral ear droop on behaviour.

Temperature and humidity affects dairy cows' willingness to be on pasture
Alfredius, Hanna[1], Norell, Lennart[2] and Nielsen, Per Peetz[1], [1]Swedish University of Agricultural Sciences, Department of Animal Nutrition and Management, Kungsängen Research Centre, 753 23 Uppsala, Sweden, [2]Swedish University of Agricultural Sciences, Unit of Applied Statistics and Mathematics, P.O. Box 7013, 750 07 Uppsala, Sweden; per.peetz.nielsen@huv.slu.se

When cattle are kept outside during the winter in Sweden the management is regulated trough the animal protection law regarding provision of shelter. This is, however, not the case during the summer grazing period. The effect of temperature humidity index (THI, mean 60.5, range 32.5 - 77.4) on dairy cows willingness to be on pasture was examined. Four years of information from Kungsängens Research farm regarding milk production and time for passing a gate between the barn and pasture of dairy cows milked in an AMS were studied. In total the production (average 26.8 kg milk/day) and behaviour data from 143 Swedish Red cows spread over lactation 1 to 6 was analysed for time spend on pasture and daily milk production in relation to THI. The cows could move freely between the barn and pasture, which were situated around the barn (between 50-260 m from the barn to the entrance of the pasture) and they were fed concentrate according to milk production in the barn. Weather data (temperature and humidity) was collected from a weather station 3 km from the barn. On average the cows spent 7:48 hr on pasture during 24 hrs. The duration spent outdoors was affected by the daily maximum THI and the time decreased with 2:25 min when THI increased with 1 ($P<0.001$). An increase in maximum temperature on its own did also negatively affect the duration the cows spent on pasture and for each 1 degree increase in maximum temperature the cows spent 2:31 min less on pasture per day ($P<0.001$). A high daily mean THI did not affect the milk production on the same day, but it negatively affected the milk production two days after with 0.026 kg when mean THI increased with 1 ($P<0.01$). These results show that even under Swedish conditions heat stress might negatively affect dairy cow and an increase in temperature and humidity might be associated to a decrease in time spent on pastures and a lower milk production. However more refined studies are needed in order to answer this question more thoroughly.

The behaviour of horses kept in tie stalls and loose boxes

Kjellberg, Linda[1] and Rundgren, Margareta[2], [1]Ridskolan Strömsholm AB, Ridskolan Strömsholm, SE-734 94 Strömsholm, Sweden, [2]Swedish University of Agricultural Sciences, Dept. of Animal Nutrition and Management, P.O. Box 7024, SE-750 07 Uppsala, Sweden; linda.kjellberg@stromsholm.com

In Sweden horses are still kept in tie stalls. However, it is not clearly documented how that affects their behaviour. The aim of this study was to compare lying behaviour in tie stalls with that in loose boxes. Four geldings and four mares (age 5-13 years) were videotaped 3× 24h in both tie stalls (length 3.0m, width 1.65 m, solid walls 1.8m in the head end) and loose boxes (3.5 × 2.9m, solid walls 2.1m, half-grid front) using a crossover design. All horses were well accustomed to both systems, bedded with wood shavings. They were kept in the box/stall for at least four days before video-taping. The behaviours (eating, standing alert, standing passive, lying, 'outside') were registered every 5th min and expressed as % of total observations. Lying duration was registered continuously and measured in minutes separating sternal and lateral recumbence. Lying down/getting up episodes were counted. Results were analysed with the programme SAS, continuous variables in proc GLM with a model including system (box, tie stall), taping day nested within system (1, 2, and 3), sex and horse within sex (1-8) as main factors. Lying down/getting up episodes were analysed with chi^{2} test. The interval registrations revealed an effect of stable in % lying behaviour (tie stall 10.1%, box 12.7%, $P=0.005$), but no differences in the other behaviours. A detailed analysis of the continuously registered lying behaviour revealed that the horses lied down in more and shorter periods in the tie stalls than in the boxes ($P<0.001$). In the boxes they also performed a rolling behaviour prior to 15% of the getting up episodes. This was only observed once in the tie stalls ($P<0.01$). The horses were mainly lying during the night and the longest lying periods lasted: tie stall 73, box 80 minutes. In conclusion, horse lying behaviour is disturbed in tie stalls.

Effect of four water resources on the behaviour of pekin ducks (*Anas platyrhynchos*)

O'Driscoll, Keelin and Broom, Donald, University of Cambridge, Department of Veterinary Medicine, Madingley Road, Cambridge CB3 0ES, United Kingdom; keelin.odriscoll@gmail.com

This study evaluated effects of four water resource (WR) treatments on water related behaviours of Pekin ducks. Ducklings (n=1,600) were randomly assigned to one of four treatments at 20 days (d) post-hatch: a chicken (CH) or turkey bell (TU), trough (TR) or bath (BA). There were four replicate groups of 100 ducklings per treatment. Treatments represented increasing levels of access to water: the beak tip in CH, beak in TU, head and beak in TR, and whole body access in BA. Behaviour was video recorded on d21, d32, d42 and d45 between 10:00 and 22:00, and scan samples taken each 7.5 min. Data were analysed using the Mixed procedure of SAS. Fewer BA ducks (6±0.2% [mean±se]) were observed near the WR than all other treatments (8±0.2%-9±0.2%; $P<0.001$), but a lower % of BA ducks (6±2%) were resting inactive at the WR (16±2%-23±2%; $P<0.001$). The % ducks resting inactive next to TR (16±2%) was also lower than CH (23±2%; $P=0.05$). However, the % ducks performing bathing behaviours (feather manipulation, head and body dipping with water) was higher in BA (46±2%) than in CH (24±2%; $P<0.001$), TU (28±2%; $P<0.001$) and TR (38±2%; $P<0.05$), and was higher in TR than in CH ($P<0.001$) and TU ($P=0.01$). Thus an increasing level of access to water appeared to promote bathing behaviour. Between 15% and 30% of bathing in BA was observed in the bath, but overall incidence of bathing was similar to TR. Ducks in CH performed a greater % of feather manipulating behaviour than the other treatments ($P<0.001$), but a lesser % of head dipping behaviour ($P<0.001$). This treatment had most limited water access, which could have inhibited or prevented head dipping. Overall, a WR that permits at least head access to water is likely to promote bathing behaviour, which is a natural behaviour in ducks, and thus has a positive effect on duck welfare.

Comparative aspects of decubital shoulder ulcers in sows: does it hurt?

Herskin, Mette S.[1], Bonde, Marianne K.[1], Jørgensen, Erik[2] and Jensen, Karin H.[1], [1]Aarhus University, Department of Animal Health and Bioscience, Research Centre Foulum, P.O. Box 50, DK-8830 Tjele, Denmark, [2]Aarhus University, Department of Genetics and Biotechnology, Research Centre Foulum, P.O. Box 50, DK-8830 Tjele, Denmark; Mettes.herskin@agrsci.dk

Decubital shoulder ulcers are lesions on the shoulders of sows kept in intensive production systems, reported to have a relatively high prevalence, and to some extend be comparable with human pressure ulcers. The lesions vary from superficial, where redness of the skin is the only clinical sign, to deep ulcers involving subcutaneous layers or even bone tissue. In sows in production systems, the ulcers are caused by pressure inflicted by the flooring, leading to oxygen deficiency in the skin and underlying tissue. In general, the presence of skin lesions, such as ulcers, is a welfare problem for farm animals reflecting that the production conditions prohibit the normal ability of the animals to adapt. At present no direct evidence for the involvement of pain in porcine decubital shoulder ulcers is available, little is known about expressions of prolonged pain in sows, and often no pain relief is given when decubital shoulder ulcers are recognised in sow production systems. However, based on the involved tissue structures, and data from human patients suffering from pressure ulcers, we suggest that both the development and presence of decubital shoulder ulcers in sows are painful and prolonged conditions. Further research is needed to confirm this – focusing on both spontaneous pain and potential hyperalgesic states – as well as to identify proper therapy.

Lying behaviour of out-wintered beef cows in Sweden

Herlin, Anders H.[1], Graunke, Katharina[2], Lindgren, Kristina[3] and Lidfors, Lena[4], [1]Swedish University of Agricultural Sciences, Rural Buildings and Animal Husbandry, P.O. Box 59, 230 53 Alnarp, Sweden, [2]Leibniz Institute for Farm Animal Biology (FBN), Research Unit Behavioural Physiology, Wilhelm-Stahl-Allee 2, 18196 Dummerstorf, Germany, [3]Swedish Institute of Agricultural and Environmental Engineering, P.O. Box 7033, 750 07 Uppsala, Sweden, [4]Swedish University of Agricultural Sciences, Animal Environment and Health, P.O. Box 234, 532 23 Skara, Sweden; anders.herlin@ltj.slu.se

The welfare of cattle kept outdoors during winter in Sweden has been questioned due to the cold weather conditions and lack of suitable lying places when it's snow, frost or very wet grounds. The lying behaviour of cattle during the winter period was investigated in herds in southern Sweden during two winter seasons. The aim was to show the variation of the lying behaviour in beef cattle kept outdoors during winter time and to identify factors which are most important. In this preliminary study, 29 dry beef breed cows aged 4 to 12 years old were studied. Lying and standing were recorded every 15 minutes by activity sensors (IceTag3D, IceRobotics, UK) which were put onto the left hind leg of seven or eight cows at the time. Recordings were made in two periods of 20 days in each herd. Average temperature was 0.8 °C (SD 4.9, with minimum and maximum average day temperatures of -11.9 and 8.9 respectively. Wind speed was on average 2.8 m/s (SD 1.5, maximum 7.6).Lying percentage per cow and day was calculated and a stepwise regression was made (Minitab 15, USA). Cows spent lying 41% (SD 9) per day on average. Lying average varied between 31-53% among cows. Minimum lying per day differed between 8-35% and the maximum between 45-71% among cows. The range of minimum and maximum lying within cow differed from 21-45%. Stepwise regression showed significant effects of days ($P=0.039$) and periods ($P=0.048$) but not for herds and cows. Less lying was seen on days with lower temperatures but days with lying below 30% were seen on days with temperatures between +4.5 and -3.9 and not for the coldest days. There was a trend towards lower lying on days with higher wind speeds but there was a large variation.

Repeated gentle handling in beef cattle: heart rate and behaviour

Schulze Westerath, Heike[1], Probst, Johanna[2], Gygax, Lorenz[3] and Hillmann, Edna[1], [1]Instiute of Plant, Animal and Agroecosystem Sciences, ETH Zurich, Universitaetsstrasse 2, 8092 Zurich, Switzerland, [2]Research Institute of Organic Agriculture FiBL, Ackerstrasse, 5070 Frick, Switzerland, [3]Centre for proper housing of ruminants and pigs, Federal Veterinary Office, Taenikon, 8356 Ettenhausen, Switzerland; heike-westerath@ethz.ch

A good animal-human relationship is one important aspect concerning cattle welfare. The aim of this study was to investigate the effect of gentle handling at head and neck on behaviour and heart beat parameters in beef cattle (seven heifers, five bulls, 8.5-11.5 months old). On each of 5 days (over 7 weeks) handling was applied twice for 4min, with 20min between daily sessions, with the animals fixed in the feeding rack. The handler was unfamiliar for the animals before the study. Heart beat parameters were recorded via Polar® system from 4min before until 15min after each handling. Data were analysed using generalised linear mixed-effects models. With increasing number of handling days, more animals reacted positively to the handling indicated by stretching of the neck and absence of defensive behaviours, and this was more pronounced in the second handling session on a given day compared to the first (day*session, $P<0.001$; 0% each on day 1 to 55% and 81% on day 5, respectively). Heart rate was slightly lower after handling (73.5 ± 1.1 before, 72.6 ± 1.1bpm after; $P<0.01$). It also decreased with increasing number of handling days (71.9 ± 1.2 on day 1 to 69.8 ± 1.3bpm on day 5, $P<0.05$) and from first to second handling session on a given day (74.1 ± 0.8 to 72.3 ± 0.8bpm; $P<0.001$). Heart rate variability (RMSSD) was slightly higher during and after the handling (14.7 ± 0.7 and 15.1 ± 0.8ms compared to 13.8 ± 0.7ms; $P<0.01$). A decrease in heart rate during the course of test and test repetitions is interpreted as the animals calming down suggesting a kind of habituation to the test procedure. However, this effect was small. A clear appeasing effect of the handling, indicated by increased RMSSD during handling could not be detected. Changes in the perception of handling, as indicated by behaviour were not found to be reflected in the heart rate variability.

Environmental conditions do interfere with emotionality and cognitive abilities in horses: example of riding schools

Lesimple, Clémence, Fureix, Carole, Le Scolan, Nathalie, Richard-Yris, Marie-Annick and Hausberger, Martine, UMR CNRS 6552 - EthoS, Station Biologique de Paimpont, 35380 Paimpont, France, Metropolitan; clemence.lesimple@univ-rennes1.fr

Horses' behaviour in riding schools is of a wide importance regarding caretakers' and users' security. Previous studies showed that numerous intrinsic or extrinsic factors, such as breed, housing conditions, sire, and work could modulate horses' emotional and cognitive abilities. Here, we compared the behaviour of 184 horses from 22 riding schools, practicing the same type of work, but differing in particular in terms of housing conditions proposed. Three emotionality tests (the arena and the novel object tests, with the horse released alone in a known arena and faced or not with a novel object, the bridge test with the horse led by the experimenter across a mattress) and one instrumental learning test (the chest test: in the box, the horse tries to open a wooden chest) were used to characterize the schools and determine how general management could explain the potential differences between sites. For the arena and novel object tests, emotionality indices were calculated, as well as frequencies of occurrence of active locomotion patterns (trot, canter and passage). For the bridge and the chest tests, times spent to cross or open were recorded. The ANOVAs and FCA showed that riding schools could be classified amongst four categories according to the behaviour of their horses. In accordance to previous studies, two factors appeared to be particularly relevant: breed, that impacts on the time to cross the bridge (Kruskall-Wallis, $H(14, N = 184)=27.08$, $P<0.05$) and housing conditions with box housed horses showing higher levels of emotionality (MW: $P<0.05$). These results underline the importance for riding schools owners to take into account the individual characteristics of the horse, and to have a questioning about the impact of general management on horses' reactivity, and the consequences in terms of security. Other factors (feeding practices, work conditions...) remain to be tested.

Environmental enrichment: effects of complexity and novelty on stereotypic and anxiety-related behaviour in laboratory mice

Gross, Alexandra N.[1,2], Engel, A. Katarina[1], Richter, S. Helene[1,2] and Würbel, Hanno[1], [1]University of Giessen, Animal Welfare and Ethology, Frankfurter Str. 104, 35392 Giessen, Germany, [2]University of Münster, Behavioural Biology, Badestr. 13, 48149 Münster, Germany; alexandra.gross@vetmed.uni-giessen.de

Improving standard cages of laboratory mice by environmental enrichment can reduce cage stereotypies and anxiety-related behaviour in behavioural tests. However, it is unclear whether success depends on specific enrichment items, environmental complexity, or novelty associated with enrichment. The aim of this study was therefore to dissociate complexity and novelty, and to study their effects on stereotypic and anxiety-related behaviour selectively. Thus, 54 freshly weaned male CD-1 (ICR) mice were allocated pairwise to standard laboratory cages enriched in three different ways (n=9 pairs per group). Condition 1 consisted of cotton wool as nesting material. Conditions 2 and 3 were more complex enrichments, including a shelter and a climbing structure as additional resources. To render complexity and novelty independent of the specific enrichment items, three shelters (cardboard house, plastic tunnel, red plastic house) and three climbing structures (ladder, rope, wooden bars) were chosen to create nine different combinations of enrichment. In condition 2, each pair of mice was assigned to a different combination that remained constant throughout the 9 weeks, whereas in condition 3, each pair of mice was exposed to all 9 combinations in turn by changing them weekly in a pseudorandom order. After 9 weeks, stereotypy levels during the first 2h of the dark phase were assessed from video recordings using one-zero sampling, and anxiety-related behaviours were assessed by video-tracking in two behavioural tests (elevated zero-maze, open-field). Mice spent up to 56% of their active time stereotyping (barmouthing). However, no differences in stereotypy scores and anxiety-related behaviours were found between the three groups (GLM, $P>0.05$, SPSS). These findings indicate that neither complexity nor novelty of enrichment had additional effects on stereotypic and anxiety-related behaviours beyond those of the nesting material.

Maternal deficit, cannibalism, infanticide or simply mortality in laboratory mouse breeding?

Olsson, I. Anna S.[1], Weber, Elin M.[1,2] and Algers, Bo[2], [1]IBMC-Instituto de Biologia Molecular e Celular, Laboratory Animal Science group, Rua Campo Alegre 823, 4150-180 Porto, Portugal, [2]Swedish University of Agricultural Sciences, Department of Animal Environment and Health, P.O. Box 234, 532 23 Skara, Sweden; olsson@ibmc.up.pt

Pup mortality is a considerable problem in laboratory rodent breeding, in particular in mice with over 40% mortality until weaning reported for some genotypes in research animal facilities. High fecundity with large litters at short intervals is a prominent feature of the reproductive strategy of rodents. Mortality both of single pups and of entire litter under certain conditions may very well be part of the normal reproductive picture. Several situations can be defined in which it would be adaptive to kill offspring: when a large litter contains small pups these may be unlikely to survive until weaning, when food is insufficient reducing litter size is adaptive, and actively eliminating an entire litter could potentially also be adaptive if conditions are such that the female is unlikely to raise offspring successfully. Nevertheless mortality, and in particular loss of whole litters, is a practical problem when breeding laboratory mice. In this review, we identify the main aspects of perinatal mortality in laboratory rodents from a behavioural perspective. Based on this, we develop a brief discussion of the use of terminology in the existing literature referring to perinatal mortality. In literature the problem is described using a range of terms of which maternal deficit, cannibalism and infanticide are predominant. These terms suggest that dam behaviour underlies or even actively causes mortality; however very rarely this is backed up with data reliably indicating that this is known. At present, very little is known about what causes pup mortality under normal housing and husbandry conditions. Against the background of present knowledge, we propose that the general term be 'perinatal pup mortality', with '(maternal) cannibalism' being reserved for those instances where there is evidence that pups are eaten and 'active (maternal) cannibalism' for those instances where the active killing of pups can be observed.

Vocalizations are reliable indicators of pain during electroejaculation in rams

Damián, Juan Pablo and Ungerfeld, Rodolfo, Facultad de Veterinaria, Universidad de la República, Departamento de Biología Molecular y Celular and Departamento de Fisiología, Lasplaces 1550, Montevideo 11600, Uruguay; jpablodamian@gmail.com

Electroejaculation (EE) is a widely used, but stressful technique in ruminants. During EE rams vocalize frequently. Our objective was to determine if vocalizations were provoked by pain, applying high epidural anesthesia to block noniceptive perception. We used 10 Corriedale x Milchschaf 2-y old rams, that were submitted to EE every 15 d since 2 months of life. During a normal EE procedure, a 23 mm diameter rectal probe with 3 ventrally oriented longitudinal electrodes was inserted into the rectum and the rams were stimulated with the same sequence of pulses (8 volts) for 3 s, and a pause of 3 s after each pulse, until 11 stimuli total had been administered. Sound emissions during EE were recorded using a InsigniaTM NS-4V24 and a microphone. The sound measurements and spectral analysis were performed with WaveSurfer 1.8.5 acquisition software. The same procedure was repeated after application of high epidural anesthesia with lidocaine 2%, 20 min. Rams vocalized 13.8±2.4 times during EE without anesthesiaThe latency from the beginning of EE to first vocalization was 2023.8±374.6 mseg. The mean duration of each vocalization, and total time vocalizing were 527.9±77.0 and 7204.1±1145.0 mseg, respectively. The fundamental frequencies (Hz) for Fstart, Fend, Fmin and Fmax were 182.2±16.3, 188.0±16.5, 172.6±15.7 and 207.7±17.8, respectively. The high values of total time of vocalizations were between 2 and 4 pulses, and the lower values were at 7 and 9 pulses ($P=0.047$, ANOVA for repeated measures). This result coincides with the pulses in which the rams ejaculated (3.0±0.4 stimulus). On the other hand, when rams were anaesthetized, no vocalizations were recorded. Vocalizations were related to nociception as were blocked after the use of high epidural anesthesia with lidocaine. Therefore, these vocalizations (with its characteristic profile) might be considered as a reliable indicator of pain and welfare during EE in rams.

Responses to short-term exposure to simulated rain and wind by dairy cattle

Schütz, Karin[1], Clark, Kate[1], Cox, Neil[1], Matthews, Lindsay[1] and Tucker, Cassandra[2], [1]AgResearch Ltd, Ruakura Research Centre, East street, Private Bag 3123, Hamilton 3240, New Zealand, [2]University of California, Department of Animal Science, 1 Shields Ave, Davis, CA 95616, USA; karin.schutz@agresearch.co.nz

Our objective was to examine how short-term exposure to wind or rain, or the combination of wind and rain, influences behavioural and physiological responses and the motivation for shelter. Twenty-four non-lactating, pregnant Holstein-Friesian cows were individually housed on rubber mats and allocated one of four treatments (control, wind, rain, wind and rain) created with fans and sprinklers. Feed intake and behavioural and physiological variables were recorded for 22h. Restricted maximum likelihood was used to assess the effects on these variables of the main effects of rain, wind and their interaction. Motivation to use the shelter was assessed by creating a trade-off between time spent feeding while exposed to the weather and time spent in the shelter, where no feed was available. Feeding times were manipulated by placing frames with three different mesh sizes over the silage; the purpose of the smaller mesh was to increase feeding time. However, shelter use was unchanged by these costs ($P=0.637$). Cows reduced their feed intake by 62% when exposed to rain and rain/wind ($P_{rain}<0.001$). Cows spent approximately 50% of their time in the shelters in all weather treatments and spent little time lying, especially under wet conditions (5.9, 4.4, 2.8, and 1.1 ± 1.4h/22h, mean±SED for control, wind, rain, and wind/rain treatments, respectively, $P=0.007$). Rain alone, and in combination with wind, decreased skin temperature (measured using infrared thermography) by 26%, on average ($P_{rain}<0.001$). The short-term response to wet conditions was characterized by a marked decline in lying time, feed intake and skin temperature. Wind alone had little effect on these responses, but magnified the effect of rain on feeding behaviour. These results indicate that protection from both rain and the combination of rain and wind is likely important for animal welfare, but future work is needed to understand when and how to best provide shelter to pastured dairy cattle.

The effects of interval of testing and quality of resource on the choice behaviour of Hy-Line Brown laying hens in a Y-maze preference test

Laine, Sonja M.[1], Cronin, Greg M.[2], Petherick, J. Carol[3] and Hemsworth, Paul H.[1], [1]Animal Welfare Science Centre, University of Melbourne, Parkville, Vic 3010, Australia, [2]Faculty of Veterinary Science, University of Sydney, Camden, NSW 2570, Australia, [3]Agri-science Queensland, Department of Employment, Economic Development and Innovation, Rockhampton, Qld 4702, Australia; s.laine@pgrad.unimelb.edu.au

The design of Y-maze tests may potentially affect animal motivation and choice, leading to spurious results. The effects of testing interval and resource quality on the choice behaviour of 11 hens were examined. Two resources, previously shown to be important to hens, were presented in the Y-maze. The dustbathing/foraging substrate offered the opportunity to test the quality of resource, in comparison to another attractive resource, social contact. Half the birds were given sawdust and other half peat moss. Hens had experience of only their designated substrate. Birds were allocated to testing-interval treatments: daily (T1), alternate (T2), or every third day (T3). Hens were trained and then tested over two, 13-day periods, during which they were deprived of both resources. Birds were allocated to different testing intervals in each period. Hens predominantly chose the substrate (88.3% substrate vs. 11.7% for social, overall). Choice was analysed as a binomial logistic generalised linear model and time to choice analysed using residual maximum likelihood mixed model analysis of the logarithm of the measurement mean. The substrate by interval interaction was examined by including the interaction term. Interval did not affect choice (transformed (back-transformed proportions of trials substrate selected): 1.52 (0.87), 2.1 (0.92), 2.6 (0.95) for T1, T2, T3 respectively (SED 0.88-1.00), $P=0.52$) and there was no interval by substrate interaction ($P=0.47$). T1 hens tended to be slower to leave the start box (transformed (back-transformed latency) for T1, T2 and T3: 0.29 (1.9s), -0.27 (0.5s), -0.08 (0.8s) respectively (SED 0.21-0.22), $P=0.06$), but there was no effect of interval ($P=0.17$), nor was there a substrate by interval interaction ($P=0.65$) on the time to choice. Thus, hens preferred substrate over social contact, but neither the testing interval nor the type of substrate influenced speed of choosing or choice. Further research is required to confirm lack of effects.

Is there any association between fluctuating asymmetry and egg shell quality in laying hens?

Prieto, María Teresa and Campo, Jose Luís, Instituto Nacional de Investigación y Tecnología Agraria y Alimentaria, Ctra. Coruña km 7.5, 28040, Spain; mtppablos@inia.es

Fluctuating asymmetry (small and random deviations from perfect bilateral symmetry) estimates developmental instability caused by environmental and genetic stress, and it is considered as an indicator of welfare and performance. On the other hand, egg shell quality has been suggested by some authors as a possible welfare indicator in laying hens. The aim of the present study was to assess the correlation between fluctuating asymmetry and egg shell quality. We hypothesized that hens with greater fluctuating asymmetry would show worse shell quality than those with smaller fluctuating asymmetry. Fluctuating asymmetry was calculated measuring five bilateral traits (middle toe length, leg length, leg width, wing length and wattle length) in 166 thirty-six-week-old hens from twelve different Spanish breeds. A total of 611 eggs were analyzed for shell quality estimated by specific gravity using the flotation method (eight different densities solutions, starting at 1.072 to 1.100 g/cm^3 in 0.004 g/cm^3 intervals). The correlation between the different fluctuating asymmetries and shell quality was calculated using residual variances and covariances. The correlations were not significant in any case, with values between 0.01 and 0.05. There were significant differences between breeds ($P<0.001$) in shell quality, and fluctuating asymmetry of middle toe length, leg length, leg width, wing length and wattle length ($P<0.001$), whereas the combined fluctuating asymmetry of all the five bilateral traits was not significant. The results do not show evidence of correlation between fluctuating asymmetry and egg specific gravity in laying hens no subjected to stress conditions, suggesting that shell quality measured by specific gravity is not a stress indicator itself.

Effects of intracerebroventricular administration of oxytocin or prolactin-releasing peptide on feeding behavior in cattle

Yayou, Ken-ichi[1], Kitagawa, Sayuki[2], Kasuya, Etsuko[1], Sutoh, Madoka[3], Ito, Shuichi[4] and Yamamoto, Naoyuki[5], [1]National Institute of Agrobiological Sciences, Lab. Neurobiology, Tsukuba, 305-8602, Japan, [2]Utsunomiya University, Utsunomiya, 321-8505, Japan, [3]National Institute of Livestock and Grassland Science, Tsukuba, 305-0901, Japan, [4]Tokai University, Aso-gun, 869-1404, Japan, [5]National Agricultural Research Center for Western Region, Ooda, 694-0013, Japan; ken318@affrc.go.jp

In rodent species, there are plenty of evidences that oxytocin inhibits feeding behavior by its central action. One of several factors mediating the satiating function of oxytocin, prolactin-releasing peptide (PrRP) is an important mediator to relay satiety signaling from gastrointestinal tract to central oxytocin neurons. In highly producing dairy cattle, a substantial dip in voluntary feed intake takes place from late pregnancy to early lactation when central oxytocin function is enhanced for parturition and lactation, which contributes to metabolic and infectious disease in early lactation. To examine the possible involvement of central oxytocin and/or PrRP in this dip in feed intake, we investigated the effects of intracerebroventricular administration of oxytocin or bovine PrRP in Holstein steers. Four Holstein steers (7-8 months old) were intracerebroventricularly administered 200 µl of artificial cerebrospinal fluid (aCSF), oxytocin (5 or 50µg), or bovine PrRP (2 or 20 nmol) 30 min before feeding. Continuous behavior sampling was performed for 60 min after the start of feeding. The catheter for intracerebroventricular administration was implanted under isoflurane anesthesia more than 1 month before the experiment. Their food ingestion got normal within a few days after the surgery, and the sign of ill-health such as fever and cough caused by surgery itself was subtle. Significant effect of treatment was not observed in any of the behavioral parameters recorded including the total number of 'interruption of feeding activity' defined as any interruption of feeding activity with the head of steers raised above the brim of feeder, drinking water, and self-grooming, and the total amount of feed intake during 60 min observation period ($P > 0.1$; Friedman's test). These results suggest that both oxytocin and PrRP might not play an important role in mediating satiating function within the brain of cattle unlike in rodent species.

Influence of the pairing system on the behaviour of red-legged partridge couples (*Alectoris rufa*) under intensive farming conditions

Alonso, Marta E., Prieto, Raquel, Sánchez, Carlos, Armenteros, José A., Lomillos, Juan M. and Gaudioso, Vicente R., University of León, Animal Production, Campus de Vegazana sn, 24071, Spain; marta.alonso@unileon.es

Millions of red-legged partridges (Alectoris rufa) are released across Europe each year for repopulation or hunting purposes, all are produced on intensive farming facilities under a forced pairing system. We studied the differences in the ethological response between partridges who were forced couples and partridges who were free couples that were kept under commercial farming conditions using two groups of 24 couples each one: (a) a group of forced pairing, one male and one female randomly chosen were introduced in the same cage and (b) group of free pairing, a female had the opportunity to choose between four males, using as female choice parameter the time spent by the female near each male. Differences in the behaviour of the couples during the first week they were together were evaluated using the ANOVA test. Female red-legged partridges chose males of higher weight ($F_{(1,94)}= 13.26$ $P<0.0001$) and these couples display more frequently patterns of feeding ($F_{(1,718)}= 54.69$ $P<0.001$), vigilance ($F_{(1,718)}= 134,81$ $P<0.0001$) and cohesive ($F_{(1,718)}= 10.491$ $P<0.001$) behaviour that increase the reproductive success of the couple. Aggressive behaviour in the free couples was significantly lower ($F_{(1,718)}= 8.80$ $P<0.001$) than in the forced couples. To have the opportunity to choose a partner diminished the agonistic behaviour incidence and increased the display of patterns of cohesive and feeding behaviour that increased the welfare of the female red-legged partridges on the farm. These results show the influence of the pairing method on the welfare of the farmed partridges when space is restricted (approximately 0.5 m^2), an effect that was not observed in previous studies with more space allowance (4 m^2).

Quality of treatment during lactation affects post-weaning behaviour of piglets

Sommavilla, Roberta[1], Hötzel, Maria J.[1], Dalla Costa, Osmar A.[2], Bertoli, Francieli[1], Silva Cardoso, Clarissa[1] and Machado Filho, Luiz Carlos P.[1], [1]Universidade Federal de Santa Catarina, Departamento de Zootecnia e Desenvolvimento Rural, Rod. Admar Gonzaga 1346, 88034-001, Brazil, [2]Embrapa, CNPSA, BR 153, km 110, 89. 7000-000, Brazil; mjhotzel@cca.ufsc.br

The aim of this study was to compare the short-term post-weaning behaviour of piglets treated either gently (GEN) or aversively (AVER) during lactation. Twenty-four lactating sows and their litters were housed in separate rooms according to treatment. A female experimenter (P1) in charge of feeding and cleaning from days 10 to 28 after birth was noisy, moved harshly and unpredictably, and shouted frequently during routine cleaning of facilities and animal handling of AVER groups. During the same routine with GEN groups she used a soft tone of voice and was careful. At weaning, an approach test was conducted with four piglets from each litter. Scores ranged from 1 (experimenter could touch piglet) to 4 (piglet escaped vocalizing). The test was repeated twice, with a 1-h interval, with P1 wearing white, and a person unfamiliar to the piglets (P2) wearing blue coveralls. During tests, applied in random order, experimenters remained standing 2 m away from the piglet without speaking. Thereafter, litters from the same treatment were mixed and housed in separate rooms, balanced for gender and liveweight (n=12 groups of 4 piglets/treatment). Behaviour was registered by 2-min scans, 4h/day for 4 days. Approach test responses were analyzed using Chi-square (df=3) and effects of treatment with a mixed model for repeated measures. Approach score was higher for AVER than GEN when tested with P1 (Chi-square=15.5; $P=0.001$) but not with P2 (Chi-square=0.69; $P=0.9$). Frequencies of lying and exploring were lower ($P<0.001$), whereas escape attempts, fighting and eating were higher ($P<0.001$) in AVER than GEN groups. Feed and water intake and weight gain did not differ between treatments. We conclude that four-weeks-old piglets can discriminate a handler according to the nature of treatment received previously. Piglets treated aversively seem to have more difficulty adapting to weaning than those treated gently during lactation.

The welfare assessment of stabled horses on the basis of behavioural observation throughout the year

Anjiki, Akiko[1], Sato, Toshiyuki[1] and Ninomiya, Shigeru[2], [1]Department of Veterinary Medicine, Tokyo University of Agriculture and Technology, 3-5-8 Saiwai-cyo, Fucyu, Tokyo, 183-8509, Japan, [2]Graduate School of Agricultural Science, Tohoku University, 232-3 Yomogita, Naruko-Onsen, Osaki, Miyagi, 989-6711, Japan; 50005155003@st.tuat.ac.jp

Behavioural observation is one of the noninvasive and useful methods for welfare assessment. The objective of this study was to consider the better way to assess the welfare of stabled horses by using behavioural indicators of frustration (IF). We observed 15 riding geldings (13 Thoroughbreds, 1 Oldenburg, 1 Half-bred, aged 4 to 22 years). They were kept in individual loose boxes in the same barn. Behaviour was recorded from 12:00 to 16:30 at 3 minutes intervals for 7 days in May 2009, and 3 days in August, November 2009 and February 2010. The total percentage of time spent in bedding investigation (smelling bedding or moving it with the nose), weaving, cribbing, licking, chewing, kicking, pawing and head shaking was calculated daily for each animal and defined as IF. Pearson's correlation coefficients between the daily IF values in May and between the monthly averages of IF value were calculated. The assumptions of parametric testing had been tested. When horses led to outside of the box for a long time, these data were excluded for analysis. We found more significant correlation ($\alpha < 0.002$) in pairs of days in same week (4 pairs (r=0.97, 0.96, 0.96 and 0.88; n=6, 7, 8 and 10, respectively) per 6 pairs) than those in different weeks (2 pairs (r=0.98 and 0.87; n=6 and 10, respectively) per 15 pairs) in May, suggesting that it will be better to conduct several observations over different weeks for the welfare assessment by IF. Significant correlation ($\alpha < 0.008$) was found also between May and August (r=0.78, n=14), May and November (r=0.88, n=11), and August and November (r=0.85, n=11). High correlation of IF between months except for February indicates that there are constant individual differences in the expression of IF, but it might be influenced by unknown effects in February.

The effects of dark brooders on chick growth and feather pecking on farm

Gilani, Anne-Marie, Nicol, Christine and Knowles, Toby, University of Bristol, Department of Clinical Veterinary Science, University of Bristol, Langford House, Langford, North Somerset, BS40 5DU, United Kingdom; Anne-Marie.Gilani@bristol.ac.uk

Recent experiments with dark brooders have shown promising results in decreasing feather pecking. The main concern of farmers is the possible negative effect on growth. This was a pilot study to test those effects. An initial small scale trial was run with 2 × 100 birds housed in two pens (3.90 × 3.30m) within the farm. One group had access to a dark brooder while the other had a heating lamp. At 8 weeks group size was reduced to 2 × 20. The dark brooder consisted of a wooden panel (1.26x2.50m) fitted with heating elements and plastic fringes. The temperature of these elements was kept the same as underneath the heating lamp, while house temperature was kept at 18 °C. Both groups were brooded for six weeks. Measurements were taken when the birds were 2, 8 and 16 weeks old and included observing feather pecking and measuring body weight, mortality and feather condition. At week 2, chicks from the heating lamp group were 10g heavier than the dark brooded chicks (135 versus 125g, average of 2 × 50 chicks), although both groups were overweight. This difference in weight continued, but was not large enough to cause a significant difference between the groups ($P<0.16$). No mortality occurred. Feather pecking data was log transformed and subjected to a GLM analysis. The number of gentle feather pecks/chick/10min did not differ between groups (0.81 (dark brooder) versus 0.28 (wk2), 0.04 versus 0.34 (wk 8) and 0.24 versus 0.12 (wk 16), $P<0.99$). Severe feather pecking and feather damage were hardly observed. Though this experiment was too small to say anything about feather pecking, it showed that dark brooders have no negative effect on growth, giving the green light for further large scale on farm experiments.

Feeding behaviour in two categories of aviaries in large laying hen groups

Arhant, Christine, Smajlhodzic, Fehim, Wimmer, Andreas, Zaludik, Katrina, Troxler, Josef and Niebuhr, Knut, Institute of Animal Husbandry and Animal Welfare, Department for Farm Animals and Veterinary Public Health, University of Veterinary Medicine, Veterinärplatz 1, 1210 Vienna, Austria; Christine.Arhant@vetmeduni.ac.at

In Austria due to the ban of conventional cages an increasing proportion of laying hens is housed in large units with aviaries of two main categories: Row-systems (RS), i.e. single and independent multi-tier rows, and more complex and comparatively large Portal-systems (PS), where the caretaker has access to the upper tiers and the area under the system. A survey to assess and compare the on-farm suitability of these systems was carried out. During this study an evaluation of behaviour at the feed trough (linear feeders) was made in 44 flocks of non-beak-trimmed Lohmann-brown hens. Hen behaviour was recorded with WLAN cameras at all feedings during light hours in the front, middle and rear section of the pen on the upper and lower tier. The first 5 minutes after the feeders had started were used for analyses. At each location, hens were observed in an area of 100×40 cm alongside the trough. For each flock mean values were calculated for the number of hens at the feed trough/minute, the number of hens feeding/minute, displacements/minute and aggressive events (pecking, chasing, fights)/minute. Mann-Whitney-U-Tests were used to test for differences between systems. RS and PS did not differ ($P>0.05$) in group size (5337 ± 1151 (Mean$\pm$S.D.)), age (37 ± 3 weeks) and trough length (10.6 ± 1.4 cm/hen). In PS, compared to RS, on average one hen more was found at the feed trough (PS: 5.7 ± 1.4; RS: 4.6 ± 0.9, $P=0.004$) and displacements of hens from the feed trough occurred twice as much (PS: 0.8 ± 0.5; RS: 0.4 ± 0.3, $P<0.001$). However, there was no difference in the number of feeding hens (PS: 4.5 ± 1.4; RS: 3.8 ± 1.0, $P=0.073$) or the incidence of aggressive events (PS: 0.2 ± 0.2; RS: 0.2 ± 0.1, $P=0.334$). The design of PS seems to lead to more hens remaining in reach of the feed trough causing more frequent interruptions of feeding behaviour.

Surgical efficacy and economic viability of vasectomy as an additional strategy for dog population control

Forte Maiolino Molento, Carla[1], Madureira Castro De Paula, Patricia[1], Lago, Elisângela[1] and Felipe Paulino Figueiredo Wouk, Antônio[2], [1]Federal University of Paraná, Animal Welfare Laboratory - LABEA, Rua dos Funcionários,1540, Juvevê., CEP 80035050, Curitiba, PR, Brazil, [2]Federal University of Paraná, Department of Veterinary Medicine, Rua dos Funcionários,1540, Juvevê., CEP 80035050, Curitiba, PR, Brazil; carlamolento@yahoo.com

The surplus dog population is a challenge in terms of public health, related mainly to increased risk of zoonotic diseases and accidents, and animal welfare. The adoption of vasectomy as a surgical intervention to control birth rate in dogs, within capture/vasectomize/release programmes, may lead to the formation of a territorial reproductive barrier, due to the presence of vasectomized males within a population of stray animals. The objective of this work was to collaborate to the development of an additional strategy to control dog population, comparing two dog vasectomy techniques, assessing their efficacy in azoospermy attainment and their viability in terms of surgical time. Thirteen dogs were vasectomized, five dogs were subjected to vasectomy through inguinal access and eight dogs were subjected to a new surgical approach, employing a modification of the pre-scrotal open orchietomy technique. Twelve animals presented post-surgical azoospermy at first post-intervention semen collection; one dog did not allow post-surgical semen collection. There was difference ($P<0.05$, t test) between averages of semen volume, comparing pre-surgical (887 ± 681 µL) to first post-surgical (324 ± 376 µL) collections. Mean surgical time was 20.0 and 19.8 for each technique. We conclude that both techniques lead to azoospermy in the first post-surgical semen collection. The new technique, with a single incision and pre-scrotal approach to the ductus deferens, might be relevant for the reduction of the negative impact of vasectomy on dog welfare during the post-surgical period.

Monitoring of poultry welfare using behavioral and radiotelemetric methods

Bilčík, Boris[1], Košťál, Ľubor[1], Zeman, Michal[1,2] and Cviková, Martina[1], [1]Institute of Animal Biochemistry and Genetics, SAS, Ivanka pri Dunaji, 900 28, Slovakia (Slovak Republic), [2]Faculty of Natural Sciences, Comenius University, Department of Animal Physiology and Ethology, Mlynská dolina B-2, Bratislava, 842 15, Slovakia (Slovak Republic); bbilcik@gmail.com

Poultry welfare is getting more attention due to EU legislation which bans the use of conventional cages and replaces them with alternative housing systems. To evaluate welfare objectively it is necessary to measure and analyze wide range of variables including behavior and physiological measures. One of the approaches is the use of radiotelemetry. Radiotelemetry allows monitoring of physiological functions in freely moving animals with minimal stress and interference with animal behaviour. Depending on the system, it allows chronic measurement of parameters such as motor activity, biopotentials (ECG, EMG, EEG), body temperature, blood pressure (systolic, mean, diastolic), pH (e.g. in intestine), blood flow or respiration frequency. Radiotelemetry is perspective tool for monitoring of poultry welfare and together with behavioral observations it can help to characterize the effects of various conditions relevant from the welfare point of view in laying hens and meat type chickens. Existing applications include e.g. monitoring of body temperature and heart rate to assess the stress during the transportation of broilers to slaughterhouse, estimation of the effect of sudden changes in management on laying hens, or the effect of adding perches in cages. Examples (using Data Sciences International telemetric system)from ongoing study of laying hens welfare will be given, focusing on the effect of feather pecking on heart rate, blood pressure and body temperature and individual differences between animals with various propensity to feather peck.

Can sow vocalization and good piglet condition prevent piglets from being crushed?
Melisova, Michala[1], Illmann, Gudrun[1], Andersen, Inger Lise[2], Vasdal, Guro[2] and Bozdechova, Barbora[1], [1]Institute of Animal Science, Department of Ethology, Pratelstvi 815, 104 00 Prague Uhrineves, Czech Republic, [2]Norwegian University of Life Sciences, Department of Animal and Aquacultural Sciences, P.O. Box 5003, 1432 As, Norway; melisovam@centrum.cz

This study focused on sow vocalization and piglet condition in loose housing system (farrowing pen measured 8.9 m^2) which might be both linked to piglet crushing. The probability of crushing is higher when piglets are present in the area where the sow lies down (piglets in the danger zone). The sow vocalization may decrease the probability of piglets being present in the danger zone. Furthermore, if piglet is in a good condition (i.e. heavy and not hypothermic), the chance of getting away from a near-crushing event is larger than for piglet in a bad condition (i.e. low body weight and hypothermic). We predicted that: (1) sow vocalization before lying down should decrease the number of piglets in the danger zone and the number of piglets being trapped, (2) piglets present in the danger zone have lower weights and rectal temperatures than the litter mates.18 loose-housed sows were video-recorded during Day 1 and Day 3 pp. Frequency of sow vocalization was calculated for 2 min before the sow lies down (n = 261 lying down events). Piglet weights and rectal temperatures were taken at Day 1 and 3 (n = 240 piglets). A generalized model (GENMOD) was applied to test our predictions. (1) Sow vocalization was present in 54% lying down events. Contrary to our prediction sow vocalization did not significantly affect the number of piglets in the danger zone and piglet trapping. (2) Piglets with a higher weight were observed more in the danger zone on Day 1 ($P<0.05$). On Day 3 there was no significant effect of weight on piglet position. Piglet rectal temperature did not affect piglet position. In conclusion, sow vocalization before lying down does not seem having any effect on moving piglets out of the dangerous position and on getting trapped. Piglet condition was not detected as a crucial factor for piglet position.

Is there a causal relation between individual stress reactivity, postweaning stress and the development of stereotypies in mice?

Engel, Anna Katarina[1], Gross, Alexandra N.[1,2], Richter, S. Helene[1,2] and Würbel, Hanno[1], [1]University of Giessen, Animal Welfare and Ethology, Frankfurter Strasse 104, 35392 Giessen, Germany, [2]University of Münster, Behavioural Biology, Badestrasse 13, 48149 Münster, Germany; Anna.K.Engel@vetmed.uni-giessen.de

Evidence indicates that stress during ontogeny is causally involved in stereotypy development in captive animals. Stress depends on an interaction between stress reactivity and exogenous stressors. To elucidate the role of stress in stereotypy development further, we used a strain of mouse that is particularly prone to stereotypies (CD-1 (ICR)) and exposed 18 dams and their litters to either brief (15min) or long (4h) daily mother-offspring separations or left them completely undisturbed from day 1 to 13 to induce variation in the offspring's stress reactivity. After weaning at 21 days, 4 female offspring of each dam were housed in pairs, one being exposed to chronic mild stress (weekly change of cage mate) and the other to standard housing conditions to induce variation in exogenous stressors. At 16 weeks of age, stereotypic behaviour was scored from video recordings for the first five minutes every 30 min throughout the 12h dark phase (120min per animal). Stress reactivity was assessed based on changes in plasma corticosterone in response to novelty exposure. There was no main effect of postnatal manipulations on stress reactivity (GLM, $P>0.1$), but postweaning housing and the interaction between the two treatments tended to affect stress reactivity (GLM, $P<0.1$), indicating that treatment combinations produced subtle variations in stress reactivity. However, against our prediction that higher stress exposure should be associated with higher stereotypy levels, neither postnatal manipulations (GLM, $P>0.1$), nor postweaning housing (GLM, $P>0.1$) or their interaction had significant effects on stereotypy levels. Thus, variation in stress exposure was either too weak to produce significant effects on stereotypy development, or stereotypy development in CD-1 (ICR) mice is not causally related to stress during ontogeny.

Nest building behavior and its effect on maternal behavior of sow and piglet production

Chaloupková, Helena[1], Illmann, Gudrun[1], Neuhauserová, Kristýna[1] and Šimečková, Marie[2],
[1]Institute of Animal Science, Department of Ethology, Přátelství 815, Prague - Uhříněves, 10400,
Czech Republic, [2]Institute of Animal Science, Biometric Unit, Přátelství 815, Prague - Uhříněves,
10400, Czech Republic; chaloupkova.helena@vuzv.cz

Nest building is an important part of the maternal behaviour in domestic pigs. The aims of the study were to assess the effect of nesting material on sow behaviour 24 h before and after birth of the first piglet (BFP) and the relationship between pre-partum nesting behaviour, post-partum maternal behaviour and piglet production. Sows, housed in enriched crates allowing walking around, were divided into two groups: sawdust (N=9) and straw (N=13). Sawdust was provided during the pre-parturient period; after parturition it was removed and straw was given. The nesting behaviour 24 h before and during parturition was analyzed. Posture changes, udder access were analyzed 24 h after BFP during 3 time periods (parturition, at end of parturition-12 h after BFP and 12-24 h after BFP) and nursing except for parturition. Piglet weight-gain and mortality were estimated 24 h after BFP. Data were analyzed using PROC GLM, MIXED and the probability of the piglets' mortality using PROC GENMOD in SAS. Nesting material did not affect sow pre-partum nesting behaviour and duration of parturition. Sawdust group had less nesting during parturition ($P<0.05$), less postural changes during parturition ($P<0.001$) and shorter nursing times ($P<0.05$). Longer pre-partum nesting period was associated with less nesting during parturition ($P<0.05$) and tendency for shorter parturition ($P<0.1$). Higher pre-partum nesting records were associated with longer parturition ($P<0.05$) and lower piglet weight-gain ($P<0.05$). The results suggest that sawdust can be suitable nesting material when straw is not available. Longer pre-partum nesting seems to be an indicator of good maternal behaviour, whereas more intensive pre-partum records may be an indicator of possible problems with piglet survival. Further research is needed determine whether there is an association between the intensity of nest-building behaviour and later maternal behaviour.

Diet form dramatically alters motivation, enrichment device use and podomandibulation time in captive orange-winged Amazon parrots (*Amazona amazonica*)

Rozek, Jessica, Danner, Lindsey, Stucky, Paul and Millam, James, University of California, Davis, Departments of Animal Science and Chemistry, 1 Shields Avenue, Davis, CA 95616, USA; jessicarozek@gmail.com

Time-compressed video recording, weighted feeder lids, and cages equipped with interruptible infra-red beams placed at key locations (feeder, water fount, perch) were employed to finely examine the activity budgets and foraging/ingestive behavior of Orange-winged Amazon parrots fed different diet forms, N=10. When fed a conventional pelleted diet (mean pellet size ~0.18 g; 'small' size, maintenance diet, Roudybush, Incorporated), parrots spent nearly all of their time perching, moving off of the perch ~25 times/day to retrieve pellets and 2-3 times/day to drink. In contrast, parrots fed pellets extruded to be 20-30X larger ('over-sized' pellets) showed comparable meal patterning, but time spent manipulating pellets with beak and foot before swallowing (i.e., podomandibulation time, determined by video observation) increased more than 500% (P=0.003; all data herein met assumptions for parametric analyses which were done using SAS programs, e.g., glimmix, mixed effects, or t-test). Moreover, in choice preference trials, parrots strongly preferred over-sized pellets to regular pellets, retrieving them approximately seven-times as often (P=0.0006). Similarly, when only regular pellets were offered, parrots removed 47.6 ±6.4 g/day (mean±SE), but this amount dropped to 6.5± 2.0 g/day, when over-sized pellets were offered concurrently (P=0.0001). Enrichment device use (wooden cube destruction) was dramatically reduced when over-sized pellets were available: during a 3-day period, parrots offered both regular and over-sized pellets reduced wooden cube mass by ~0.13 g, while removal of the over-sized pellet option increased mass loss 50-fold (P=0.0018). Finally, motivation tests employing weighted feeder lids, found that parrots repeatedly lifted up to 500 g (more than their own body weight) to retrieve over-sized pellets, even when regular pellets were concurrently freely available. Taken together, these results demonstrate that diet form – not nutrient content – is highly motivating and dramatically alters activity budgets, supporting the view that providing naturalistic foraging opportunities enhance welfare.

Comparison between visual scoring system and five other methods to evaluate sow gait and lameness

Grégoire, Julie[1,2], Bergeron, Renée[3], D'allaire, Sylvie[4], Meunier-Salaün, Marie-Christine[5] and Devillers, Nicolas[1], [1]Agriculture and Agri-Food Canada, Dairy and Swine R&D Centre, P.O. Box 90, J1M 1Z3 Sherbrooke, QC, Canada, [2]University Laval, Department of Animal Science, Pavillon Paul-Comtois, G1V 0A6 Québec, QC, Canada, [3]University of Guelph, Alfred Campus, P.O. Box 580, K0B 1A0 Alfred, ON, Canada, [4]University of Montreal, Faculty of Veterinary Medicine, P.O. Box 5000, J2S 7C6 St-Hyacinthe, QC, Canada, [5]INRA, SENAH, Domaine de la Prise, 35590 St-Gilles, France; Julie.Grégoire@agr.gc.ca

The strong impact of lameness on the productivity and longevity of sows bring out the need to identify and measure locomotion disorders on commercial farms. This study aims to compare a 5-degree visual scoring system of lameness with five different methods to evaluate gait and quantify lameness in sows: a foot lesion grid, a standing up test, kinematics, footprint analysis and automated recording of posture and steps. Data were collected on 50 sows of different parities and stages of gestation. Comparisons were made between the first three degrees of the lameness scoring system where sows can still walk: lame (L), slightly lame (SL) and not lame (NL), using the mixed and logistic procedures of SAS. Kinematics analysis showed that lame sows walked more slowly (L: 0.83± 0.04; SL: 0.94±0.03; NL: 0.96±0.03 m/sec; $P<0.05$) and tended to have a shorter stride length (95±2; 101±1; 98±1 cm; $P<0.06$) and a longer stance time (2.78±0.08; 2.56±0.07; 2.57±0.06 msec; $P=0.08$) than SL and NL sows. Lame sows also stepped more around feeding (10.1±1.1; 6.1±0.8; 5.4±0.8 step/min; $P<0.01$), stayed standing less time after feeding (33.0±4.1; 41.8±3.2; 48.6±3.0 min; $P=0.01$) and spent less time standing over 24h (7.6±2.9; 10.4±2.2; 17.1±2.1%; $P<0.05$) than SL and NL sows. Foot lesion recordings showed that the number of feet affected on the white line increased lameness risk ($P<0.05$). Finally, standing up test showed that L and SL sows refused more often to stand up than NL sows ($P<0.05$). No clear differences were found between sows using footprint analysis. In conclusion, kinematics analyses and automated recording of posture and steps are the most promising tools for discriminating lame sows from sound ones. However, these methods need to be adapted and validated for on-farm use.

Feeding behaviour and body temperature of newborn dairy calves in relation to feed level and heat supply

Vasseur, Elsa[1], De Passillé, Anne Marie[2] and Rushen, Jeff[2], [1]University of British Columbia, Faculty of Land and Food Systems, 2357 Main Mall, Vancouver BC V6T 1Z4, Canada, [2]Agriculture and Agri-Food Canada, Pacific Agri-Food Research Centre, 6947 Highway 7 P.O. Box 1000, Agassiz BC V0M 1A0, Canada; vasseur.elsa@gmail.com

Low ambient temperature can increase mortality of calves but we know little of calves' abilities to thermoregulate. To better understand how newborn calves thermoregulate, we examined the effect of an external heat source and feed level on feeding behaviour and body temperature in a 2 by 2 factorial design using GLM. We housed 48 Holstein calves for 4 days after birth in individual pens with a heat (HL) or a cold lamp (CL). Calves were fed 4 L of colostrum soon after birth (day 1) and provided continuously cold milk at HIGH (ad libitum) or LOW (10% BW) level from a teat bucket thereafter. Body temperature was recorded continuously on day 1 to 4 using a data logger inserted into the vaginal cavity. On day 2, calves with heat supply drank significantly ($P<0.05$) more milk (LS-Mean±SE, HL: 2.48±0.25 L vs. CL: 1.59±0.26 L) but no effect of feed level or interaction between heat supply and feed level was found on milk intake. No effect of heat supply nor feed level was found on milk intake on day 3 ($P>0.1$). On day 4, calves on HIGH diet drank significantly ($P<0.001$) more milk (LS-Mean±SE, HIGH: 6.47±0.32 L vs. LOW: 4.18±0.31 L) but no effect of heat supply or interaction was found on milk intake. On day 1 to 4, no effect of heat supply nor feed level was found on vaginal temperature (Mean±SD, day 1: 38.2±0.4 °C; day 4: 38.6±0.3 °C). Heat supply seems to help newborn calves drinking more milk but the effect of the heat lamp does not persist after 2 days of age. Newborn calves up to 2-day old have difficulty drinking large volumes of milk from a milk bar feeder. Assistance is required to insure sufficient feed intake during first 2 days.

Behavioural characteristics of sows that crush their piglets on small farms

Suzuki, Haruka and Takeda, Ken-ichi, Shinshu University, Faculty of Agriculture, 8304 Minamiminowa, Kamiina, 399-4598 Nagano, Japan; haatti3@hotmail.com

Although the management of small farms has improved, sows crushing their piglets is still one of the main causes of piglet death. The behaviour of sows in a farrowing crate can greatly influence piglet survival. This study investigated the effect of sows' behaviour by comparing the behaviour of sows that had crushed any piglets (Cr) and had not (NCr) before and during the study. Two video cameras continuously recorded three Cr (mean parity 5.0±2.6, litter size 10.3±2.3) and five NCr (mean parity 4.3±2.1, litter size 10.3±4.9) Large White × Landrace sows housed in farrowing crates for one week before farrowing from the day before parturition until 48 h after the first piglet was born. Behaviours were categorised as lateral lying (L), lying on the belly (B), dog sitting (D), and standing (S), and their frequencies and duration were recorded at one-minute intervals. Transitions in posture were also noted from observations and the times spent shifting were measured. There were no significant differences in the frequency or duration of each behaviour between NCr and Cr sows. Cr sows changed posture from S-D more frequently (Cr: 50.3±37.5 vs. NCr: 2.8±5.2, t-test, $P<0.05$) and D-S (73.7±26.6 vs. 36.8±14.8, $P= 0.05$) and less frequently from L-D (14.0±5.2 vs. 26.8±8.5, $P<0.05$) than NCr sows. NCr sows changed posture fewer times between 24 h before parturition and 24 h after (17.1±18.8 vs. 3.9±4.7, paired t-test, $P<0.01$) and 24 to 48 h after (3.9±4.7 vs. 2.4±3.0, $P<0.01$); Cr sows did not change posture significantly between 24 h before and after (12.6±16.5 vs. 10.2±13.4). Cr sows spent less time lying or sitting down than did NCr sows (11.0±4.0 vs. 18.4±4.4, t-test, $P<0.05$). These results indicate that sows at risk of crushing piglets show some differences in behaviour.

Attitudes of consumers towards animal welfare products in Japan

Takeda, Ken-ichi[1], Fukuma, Miho[1], Nagamatsu, Miki[2] and Abe, Naoshige[3], [1]Shinshu University, Faculty of Agriculture, 8304 Minamiminowa, Kamiina, 399-4598, Nagano, Japan, [2]Nippon Veterinary and Life Science University, Department of Animal Science, 1-7-1, Kyonancho, Musashino, 180-8602, Tokyo, Japan, [3]Tamagawa University, College of Agriculture, Department of Bioenvironmental Systems, 6-1-1, Tamagawagakuen, Machida, 194-8610, Tokyo, Japan; ktakeda@shinshu-u.ac.jp

In Japan, efforts to formulate housing guidelines to promote farm animal welfare began in 2009. To promote animal welfare, consumers' acceptance for animal welfare products in market is important as well as improvement housing systems in farms. This study investigated consumer awareness of animal welfare and whether consumers in Japan are prepared to pay more for animal welfare products (AWP). Non-student female consumers older than 20 years in metropolitan and rural areas were asked to complete a questionnaire. The questionnaires were distributed to consumers by mail. Consumers were asked 35 questions, including questions about their concern for animal welfare, their eagerness to buy and willingness to pay for AWP, and sociodemographic information (age, income, resident area, whether they had visited a farm, etc.,). Before replying, the participants were made to read a definition of animal welfare and an explanation of housing systems for dairy cows and finishing pigs. We used Spearman's rank correlation test (SPSS statistical package) to compare the proportion of the respondents. Of the 318 questionnaires started, 281 were completed (88.4% completion rate) of which 52% were by metropolitan residents and 48% were by rural residents. Of the respondents, 82.9% did not know animal welfare before this survey, although 80.1% were interested in animal welfare. Consumers who were concerned about housing systems for farm animals were more eager to buy AWP than those who were not concerned ($P<0.01$). There was a positive correlation between eagerness to buy AWP and rated willingness to pay for AWP ($rs=0.44$, $P<0.01$). Resident areas did not influence any of the responses. Consumers in their 40s and 50s were more eager to buy animal welfare products than those in their 20s or over 60 ($P<0.05$). Higher-income consumers were more willing to pay for AWP ($rs=0.13$, $P<0.05$).

Prenatal stress impairs maternal behaviour in the domestic pig

Rutherford, Kenneth MD, Donald, Ramona D, Robson, Sheena K, Ison, Sarah H and Lawrence, Alistair B, SAC, Animal Behaviour & Welfare, West Mains Road, Edinburgh, EH9 3JG, United Kingdom; kenny.rutherford@sac.ac.uk

Prenatal stress can have a variety of effects on biology. In farmed animals these effects can be detrimental to animal welfare. Previous work in pigs suggested that one consequence of prenatal stress (PNS), generated through social mixing of sows, was an increased likelihood of piglet-directed aggression when PNS offspring themselves farrowed. Here the behaviour of PNS or control gilts in either a farrowing crate (PNS: n=14; CON: n=7) or loose farrowing pen (PNS: n=13; CON: n=4), was continuously observed for 24hours before and after the start of parturition. Data was analysed by REML. In the 24hrs prior to the first piglet birth, gilt behaviour was significantly affected by environment but not by prior stress history (or the interaction between the two).Gilts in pens showed more fixture/substrate-directed behaviour (W=21.70; $P<0.001$) and stood up more (W=33.21; $P<0.001$) than those in crates. In the 24hours after the start of farrowing, PNS gilts spent more time ventral lying than control gilts (W=9.11; $P=0.015$). PNS gilts also spent more time focussing attention to their piglets (W=7.31; $P=0.022$) and were more reactive when piglets approached their head (W=7.65; $P=0.021$). Ventral lying and heightened reactivity towards piglets have previously been shown to be two key components of poor maternal care in pigs. Farrowing environment did not impact on the time spent directing attention to piglets (W=2.32; $P=0.14$) or piglet approach response (W=0.12; $P=0.73$). In summary, gilt maternal behaviour was negatively affected as a consequence of the mother's experience as a fetus more than a year earlier. Specifically, exposure to prenatal stress increased negative reactions to the experience of farrowing and piglet contact, irrespective of the degree of behavioural restriction experienced by mothers during the peri-parturient period. Overall, this result further demonstrates the long-lasting effects that prenatal stress can have on farm animal behaviour and welfare.

Changes in sexual behaviour in Saint Croix rams after joining with an unknown ram

Lacuesta, Lorena[1], Sánchez-Dávila, Fernando[2] and Ungerfeld, Rodolfo[1], [1]Universidad de la República-Facultad de Veterinaria, Departamento de Fisiología, Lasplaces 1550, Montevideo 11600, Uruguay, [2]Universidad Autónoma de Nuevo León, Facultad de Agronomía, Pedro Mtz. 506, Marín, Nuevo León, Marín 64000, Mexico; lacuesta16@gmail.com

The presence of dominant rams affect the sexual behaviour of subordinate rams. However, it is not known if sexual behaviour is affected during the establishment of dominance-subordination relationships. Therefore, the objective was to determine if rams' sexual behaviour toward oestrous ewes is affected immediately after joining with an unknown ram. The experiment was performed during the breeding season in Nuevo León, México ($40°$ N) with 10 Saint Croix adult rams (2-2.5 years: 67.9 ± 1.6 kg; mean $\pm$ SEM) that have been maintained in individual pens for 1 month. In Day 0, two rams of similar body weight were joined in a new pen, in which dominance relationship were determined by food competition daily during 10 days. Sexual behaviour of subordinate (SR) and dominant (DR) rams towards oestrous ewes was determined before (BJ: Days -8, -5 and -2) and after joining (AJ: Days 3, 8, 10, 14 and 17). Rams were individually exposed to an oestrous ewe for 20 min in another pen, and the frequency of courtship behavioural events, mounts and mates was recorded. The recorded parameters were compared with an ANOVA for repeated measures. Sexual behaviour of SR and DR was similar ($P>0.05$). However, most behaviours increased after joining: mount attempts (0.8 ± 0.5 vs 5.6 ± 3.1, $P<0.05$), mounts without ejaculation (2.1 ± 1.0 vs. 6.5 ± 3.1, $P<0.005$), total mounts (3.3 ± 1.2 vs 9.3 ± 3.0, $P<0.001$), and the mates/total mounts ratio (0.3 ± 0.1 vs. 0.4 ± 0.1, $P<0.001$), BJ and AJ respectively. Anogenital sniffings (15.0 ± 3.0), flehmen (2.6 ± 0.8) and mounts with ejaculation (2.1 ± 0.5) were not affected. The number of lateral approaches decreased after joining (60.0 ± 12.2 vs. 57.0 ± 14.0, $P<0.05$). We concluded that after joining unknown rams there is an increase in rams' sexual behaviour when tested alone with oestrous ewes. This change in sexual strategy may be due to the lack of competition during sexual tests.

Risk-factors for stereotypic behaviour in horses in Western and English stables in Belgium

Vervaecke, Hilde[1,2], Santamaria-Alonso, An[2], Laevens, Hans[2], Overmeire, Inge[2] and Lips, Dirk[1,2], [1]KULeuven, Centre for Science, Technology and Ethics, Kasteelpark Arenberg 30, 3001 Leuven, Belgium, [2]KaHoSL/Ass. KULeuven, Agro-& Biotechnology, Hospitaalstraat 21, 9100 Sint-Niklaas, Belgium; hilde.vervaecke@kahosl.be

The development of stereotypic behaviour in horses is known to relate to a number of factors such as breed type, feeding regime, housing and management conditions. We investigated the prevalence of these risk-factors for 217 horses in 21 Western stables (n=109 horses) and 20 English stables (n=108 horses) in Belgium. In each stable, five horses were selected at random and age, sex, breed type, details of feeding regime, housing and management conditions were scored. In addition, this was scored for the other horses that were reported to perform stereotypic behaviour. To test which horse or stable characteristics were associated with the presence of stereotypic behaviour, univariable GEE models were fitted (ranch ID was included as random effect). Risk factors with $P<0.25$ were included in a multivariable GEE model. Stable type (Western versus English) was forced into the multivariable analysis because it was of special interest. In the random sample, 10% of the horses showed stereotypic behaviour. In the overall sample, locomotory stereotypes (e.g. weaving, box circling, box kicking) were more frequent (64%) than oral stereotypes (wind-sucking and cribbing) (36%). The age-group 5-9 years showed most stereotypic behaviour ($P<0.0001$). In Western stables the horses tended to perform less stereotypic behaviour ($P<0.0550$).Western stables differed significantly from English stables in breed type, private or public use of the horses, outdoor, size of boxes, weekly frequency of movement, kind of movement, social contact, duration and frequency of hay availability. In English ranches, there were more Belgian Warmblood horses and more horses for public use. Although the horses in English stables had larger stables, there was less opportunity for social contact, fewer opportunity to walk and exercise, and they spent less time eating roughage. This means that, with the exception of box size, the risk factors for the development of stereotypic behaviour were more prevalent in English versus Western stables.

Can observed horse behaviours predict rideability scores in Belgian Warmblood Stallions?

Janssens, Steven[1], Schaerlaeckens, Sarah[2], Vincentelli, Charlotte[2], Laevens, Hans[2], Lips, Dirk[2,3] and Vervaecke, Hilde[2,3], [1]KULeuven, Livestock Genetics, Department of Biosystems, Kasteelpark Arenberg 30 - bus 2456, 3001 Heverlee, Belgium, [2]KaHoSL, Ass. KULeuven, Agro- & Biotechnology, Hospitaalstraat 21, 9100 Sint-Niklaas, Belgium, [3]KULeuven, Centre for Science, Technology and Ethics, Kasteelpark Arenberg 30, 3001 Leuven, Belgium; Steven.Janssens@biw.kuleuven.be

Rideability is an important asset for competition horses. It is a difficult to define composite measure, that tends not to be operationally defined in most contests. In a basic attempt to relate this concept to specific horse behaviours, we compared the rideability scores of jury and riders with observed behaviours at an official contest. Twenty one Belgian Warmblood Stallions, aged between three and four years that had been in training for about five months, were scored for rideability on an event where they are evaluated as future breeding stallions. The horses were scored by two new riders during a ridden test on a ten-point-scale (1= lowest, 10= highest). Simultaneously, a professional jury (off horse) of one person gave a score during the ridden test. There was no operational definition of the term rideability, it was intuitively scored by each person how good the ridden performance of the horse was, the traditional system of scoring breeding stallions in Belgium. The three scores for rideability were averaged and then compared to the set of behaviours that were observed in at least three horses (examples of omitted behaviours: rearing, tongue out, whinnying) scored by filming the horses performing the three gaits, on both sides, during three minutes. In the stepwise regression analysis with average score as dependent variable and sum of the different horse behaviours as independent variables, a model emerged (Adjusted R-square=0.55, F=11.03, $P<0.0007$) with keeping the head above the bit (beta=-0.38, $P<0.001$) and head shaking (beta= -0.07, $P<0.003$) as significant variables. Lower rideability scores related to higher scores of head shaking and walking with the head above the bit. Other behaviours such as tail-swishing or chewing related poorly to the scores. In this select sample of potential breeding stallions, movement and position of the head related best to the rideability scores of jury and riders.

Management of behavioral abnormality in singly housed Japanese macaques by movie presentation

Ogura, Tadatoshi[1,2], [1]Japan Society for the Promotion of Science, Kouji 5-3-1, Chiyoda, Tokyo, 102-8472, Japan, [2]Kyoto University, Primate Research Institute, Kanrin 41, Inuyama, Aichi, 484-8506, Japan; ogura@pri.kyoto-u.ac.jp

Some biomedical research and zoo management protocols require that nonhuman primates be housed individually, although many nonhuman primate species are highly social animals. Especially in this condition, movie presentation is a popular environmental enrichment technique. Nevertheless, there are few studies which have evaluated the effect of movie presentation on the welfare of captive primates. Moreover, the effects of contents and controllability of movies are still unclear, although visual preference of primates and positive effects of controllability over environment have been suggested. This study examined the effect of movie presentation on abnormal behavior of four Japanese macaques, Macaca fuscata, (two males and two females, 10-13 years old) which were single-housed for more than one year. An LCD monitor was placed in front of the cage of the subject and the movies were presented for an hour per day. In this setting, three contents of movies, i.e., conspecifics, human, and animation and two presentation procedures, i.e., controllable presentation and successive presentation were tested. Behaviors of the subjects during these conditions were recorded using instantaneous sampling with a 30-second interval and compared with the behaviors during control conditions in which no movie presentation was supplied. Frequencies of abnormal behavior of three subjects were significantly lower during movie presentation than during the control conditions ($P<0.05$), although one subject showed no significant difference. The three subjects showed preference for the movie of conspecifics, which differentially affected the behavior of the subjects. These results suggested that movie presentation is a useful technique to improve the behavioral abnormality of singly housed Japanese macaques. The contents of movies influence the effectiveness of the movie presentation. To assess the management of whole day life by movie presentation, the evaluation of total time budget including the period when the movie is not present in a day might be necessary.

Can broiler chickens run fast?

Ito, Shuichi[1], Koga, Yurika[1], Miura, Risa[1], Saito, Akari[1], Goto, Nodoka[1], Hagiwara, Shintaro[1], Okamoto, Chinobu[1] and Yayou, Ken-ichi[2], [1]Tokai University, Agriculture, Kumamoto-ken, Aso-gun, Minamiaso-mura, Kawayou, 869-1404, Japan, [2]National Institute of Agrobiological Sciences, Laboratory of Neurobiology, Ibaraki-ken, Tukuba-shi, Ikenodai 2, 305-8692, Japan; shuichi_ito@agri.u-tokai.ac.jp

In broiler chicken farming, low stocking density is required for a 'freedom to express normal behaviour'. However, there are some reports that chickens showed few locomotor activity regardless of stocking density at the end of the growing period. To answer the question of whether fast growing broiler may not have motivation to move or they might lose their ability to move, we estimate the locomotor ability of broiler chicken by measuring a running speed. We tested 6 Chunky broiler cocks every two days from 5 to 59 days old. They were forced to run at treadmill speeds from 1.5km/hr to 16km/hr with increment or decrement of 0.5 km/h. Once the chicken could continue running for about ten seconds at a set speed, we judged that the chicken could run at that speed, until reaching the speed set for that treadmill. A cage mate was set in front of the treadmill, and the experimenter urged an experimental animal forward by flapping paper. During the experiment period, there was no bird with any problems walking. There was an individual difference in their running ability. As for about 3,000g broiler chicken it is possible to run at an average of 8.2±1.83km/hr. The first running speed of the broiler chicken was 11km/hr, when the birds weight was from 2,000g to 3,000g. With a subsequent body weight gain, the running speed of the broiler chicken decreased. We calculated a single regression equation for each individual with maximum running speed and body weight. As a result of this study, broiler chickens at the end of the growing period have locomotor activity. Therefore, the cause of low locomotor activity of broiler chickens in this time may be affected by motivation level.

Short-term intake rate and ingestive behavior of cattle grazing dwarf bamboo (*Sasa senanensis*)

Yayota, Masato, Tani, Yukinori and Ohtani, Shigeru, Gifu University, Faculty of Applied Biological Sciences, 1-1 Yanagido, Gifu, 501-1193, Gifu, Japan; yayo@gifu-u.ac.jp

Dwarf bamboos are major native forage in Japan and other countries. Those plants are, generally, semi-woody form with stiff and fibrous leaves, but markedly reduced in size under heavy grazing. These unique forms may affect intake rate and ingestive behavior of grazing cattle, and thus animal performance. This study was to determine short-term intake rate and ingestive behavior of cattle grazing dwarf bamboo. Approximately $1.0 \ m^2$ hand-constructed swards (HCS) was prepared by fixing plants from naturally growing dwarf bamboo (Sasa senanensis) in the density of the plants $100/m^2$. Six Japanese Black cows (400 ± 49 kg) allowed to graze HCS for 5 min. The cows were fastened from 19:00 of the previous day of each session. In exp.1, each cows grazed semi-woody form (approx. 180 cm high) dwarf bamboo in 3 replicates. In exp.2, intake rate and ingestive behavior of the cows grazing semi-woody and downsized (approx. 40 cm high) dwarf bamboo were compared in a randomized block design using linear model. The intake was estimated by the difference of the pre- and post grazing HCS weight, corrected for water loss. Ingestive behavior of the cows was measured by acoustic monitoring methods. Exp.1: Estimated intake rate, chew rate and mass (mean$\pm$SE) were 36.4 ± 3.3 g DM/min, 67.2 ± 4.7/min, and 0.55 ± 0.04 g DM/chew. Bite could not clearly distinguish in a series of ingestive behavior. Exp.2: Intake rate and chew rate were not affected by the forms of dwarf bamboo ($P>0.05$), although chew mass was higher in semi-woody form than in downsized dwarf bamboo ($P<0.05$). The intake rate of dwarf bamboo was comparable with or less than that of common herbages (18.0-67.2 g DM/min) in published data.

Validation of a human approach test on dairy cows housed indoors and at pasture
Gibbons, Jenny, Lawrence, Alistair and Haskell, Marie, SAC, Animal behaviour and welfare, Sustainable Livestock Systems, West Mains Road, Edinburgh, EH9 3JG, United Kingdom; jenny.gibbons@sac.ac.uk

Human approach tests (HAT) are commonly used to assess fear of housed dairy cattle as part of on-farm welfare assessments. If a HAT is to be used more widely, then it should be applicable to a range of environments including when cows are grazing at pasture. The objectives of this study were to evaluate the reliability of HAT in two different contexts by examining the correlation between HAT carried out indoors and at pasture. Forty healthy Holstein-Friesian cows were balanced for parity and days in milk. HAT was carried out on each cow three times while housed and three times while at pasture. Flight response score was measured on an ordinal scale according to the distance at which the cow withdrew by taking two or more steps in the opposite direction from the approaching experimenter. Additionally, a qualitative assessment of differing aspects of the cow's behavioural response was scored. The experimenter marked an individual visual analogue scale for four expressive terms (at ease, nervous, shy and fearful). A low level of agreement between the median flight response score indoor and at pasture indicated low consistency within animals across the two situations (r_s=0.3, P=0.02, n=40). Cows withdrew sooner from the approaching human when at pasture compared to when they were housed (U=505.5, P=0.004, n=40). The qualitative assessment was low to moderately consistent within animals between the indoor and at pasture HAT (At ease: r_s=0.36, P=0.021; fearful: r_s=0.37, P=0.018; nervous: r_s=0.370, P=0.019; Shy: r_s=0.440, P=0.004). There were no significant differences between indoors and outdoors for at ease (T=376.0, P=0.653, n=40), nervous (T=357, P=0.839, n=40), shy (T=287.5, P=0.384, n=37) and fearful (T=306.0, P=0.178, n=40). In conclusion, the qualitative behavioural response is more reliable as a welfare assessment across environments than the flight response score.

Negative impact of feed restriction on the welfare of goat kids housed in metabolic pens

De Oliveira, Daiana, Teixeira, Izabelle Auxiliadora Molina de Almeida, Paranhos Da Costa, Mateus José Rodrigues, De Resende, Kleber Tomás, De Souza, Samuel Figueiredo, Boaventura Neto, Oscar, De Lima, Lisiane Dorneles, Figueiredo, Fernanda Oliveira Miranda and De Oliveira, Herymá Giovane, UNESP, Department of Animal Science, Via de acesso Prof. Paulo Donato Castellane, s/n, 14884900, Jaboticabal, SP, Brazil; daiana_zoo@yahoo.com.br

Feed restriction is a common practice in scientific studies and in the animal husbandry. The aim of this study was to determine the impact of the feed restriction on the behavior and welfare of pre-weaned individually penned kids. A total of 27 Saanen kids (males, females and castrates) with 33.2±3 days of life, were subjected to 3 different nutritional levels (without restriction, intermediate restriction and severe restriction), and the individual milk and solid ration intake were daily controlled. The behavioral observations were made on direct way, using video cameras on continuous recording, considering 11 categories: feedbunker interaction, water through interaction, pen interaction, reaction to human, social interaction passive, standing, lying, movement, bipedal and self-grooming, using the frequency and the length of the behaviors as variables. The experimental design was a randomized block design with factorial arrangement $3 \times 3 \times 3$ (period × nutritional level × sex). The data were analyzed as mixed linear models, using the MIXED proceeding of SAS 9.1. The dry matter intake of the solid ration was below for what was expected for kids with this age (8 g/day of DM). The social interaction of animals with humans was changed, and the kids severely restricted showed less reaction from external stimuli (average frequency 3.4; 3.9; 2.0 ocurrences/hour for animals without restriction, intermediate restriction and severe restriction, respectively, $P=0.03$), staying in standing position for longer periods, showing apathy and depression, indicating impaired welfare. All animals presented stereotypes; however the females were more sensitive to restrictive conditions, biting the pen bars more often (average frequency 15.4; 17.5; 12.4 ocurrences/hour for males, females and castrates, respectively, $P=0.04$) and for longer periods than other animals (average length 355.7; 486.2; 310.7 sec/hour for males, females and castrates, respectively, $P=0.0094$). This study showed there are costs to the animals, with alteration on the behavior and deterioration in their welfare.

Contact with estrual ewes increases testicular blood flow in rams

Ungerfeld, Rodolfo[1] and Fila, Danilo[2], [1]Facultad de Veterinaria, Deartamento de Fisiología, Lasplaces 1550, Montevideo 11600, Uruguay, [2]Facultad de Veterinaria, Deartamento de Reproducción, Lasplaces 1550, Montevideo 11600, Uruguay; rungerfeld@gmail.com

Contact with estrual ewes induces an increase in LH and testosterone concentrations in rams. LH provokes an increase in testicular blood flow in rodents,. Considering that testicular activity (endocrine and spermatogenic) is mainly regulated by hormones reaching testicle by blood, the objective was to determine if courtship and mating estrual ewes, induces an increase of testicular blood flow in rams. During the breeding season, after 2 weeks isolated from females, 18 20-min tests with one ram and one estrual ewe, allowing the rams to court and mate the ewes were performed. Blood samples were obtained before and at the end of the test. At the same moments, and 30 min later ultrasonographic examinations of the testes were performed with B-mode ultrasound scanner connected to a 7.5 MHz linear array transducer. An ultrasound digital image of each testis was evaluated with a specific software (Image-Proplus 3.01, Media Cybernetics, USA) analyzing in four spots with a diameter of 1cm on the same place of the image with the spot metering technique. The pixel intensity of the testicular parenchyma was determined in each spot (1=black to 256=white). To each ram, means and SD pixel intensity were determined for each testicle and the average of both testicles was used for statistical analysis. Rams mated 1.6 ± 0.3 times/test. Testosterone concentrations increased from 29.1 ± 3.7 to 36.6 ± 3.7 nmol/l ($P=0.016$). Pixel intensity decreased from 152.9 ± 4.7 to 132.4 ± 4.6 ($P<0.0001$), and 30 min later remained at 132.9 ± 3.1. Changes in pixel intensity were similar in rams that mated or not ($P>0.1$). The stimulus of estrual ewes determined an increase of testosterone secretion, which was associated with an increase of testicular blood flow, remaining increased at least for 30 min after the end of the ram-ewe contact. Socio-sexual stimuli, as contact with estrual females, can directly influence testicular activity through blood flow.

Systematic heterogenization improves reproducibility of behavioural test results

Richter, S. Helene[1,2], Garner, Joseph P.[3] and Würbel, Hanno[1], [1]University of Giessen, Animal Welfare and Ethology, Frankfurter Strasse 104, 35392 Giessen, Germany, [2]University of Münster, Behavioural Biology, Badestrasse 13, 48149 Münster, Germany, [3]Purdue University, Animal Sciences, 125 South Russell Street, West Lafayette, IN 47907, USA; helene.s.richter@vetmed.uni-giessen.de

Poor reproducibility of results from animal experiments has led to controversy over the logic of experimental standardization. Because standardization reduces the external validity, it may cause rather than cure poor reproducibility. We therefore tested whether systematic heterogenization instead of standardization within experiments would improve the reproducibility of behavioural test results between experiments. We allocated 256 female mice of two inbred strains (C57BL/6, BALB/c) to four standardized and four heterogenized experiments and examined them for strain differences in three behavioural tests (free-exploration, open-field, novel-object). Each experiment was characterized by a unique combination of eight experimental factors (test age, cage enrichment, cage size, test light, test time, sound level, experimenter, handling) to mimic different laboratories. Within each experiment, six of these factors were standardized, while the remaining two were either standardized or systematically heterogenized using a 2×2 factorial design. While strain differences were relatively stable among heterogenized experiments, they varied considerably among standardized experiments. Indeed, variation in strain differences among the heterogenized experiments was significantly lower (GLM, $P<0.001$), confirming better reproducibility. Moreover, we could show that this improvement was due to heterogenization increasing within-experiment variation relative to between-experiment variation (GLM, $P<0.001$). Thus, systematic variation of two experimental factors was sufficient to guarantee robust results across the unavoidable variation between experiments. Whether these findings also apply to a real multi-laboratory situation, where differences in the conditions between experiments are expected to be even greater, remains to be examined. Nonetheless, these findings show that systematic heterogenization could be a powerful tool to improve the external validity of experimental results, thereby reducing conflicting findings in the literature and preventing unnecessary duplication of experiments.

Do cows adapt quickly to being milked by robots?

Jacobs, Jacquelyn A. and Siegford, Janice M., Michigan State University, Animal Science, 1290 Anthony Hall, East Lansing MI, 48824, USA; jacob175@msu.edu

This study examined the occurrence of stress-related behaviors of cows during milking as they adapted to an Automatic Milking System (AMS). Four parameters - step-kick behavior both before and after attachment of cups, elimination (urination and defecation instances), and vocalization were measured during milking by trained observers; while milk yield was automatically recorded by the AMS. 77 cows with acceptable udder and teat conformation that would not interfere with adaptation to the AMS and that were lactating (18=early (0-100 days in milk (DIM)); 27=mid (100-200 DIM); 32=late (200+ DIM)) for the full duration of the project were chosen for observation. All cows had previously been milked in a double-six herringbone milking parlor. Data was collected for 24 hour periods beginning on the day the cows transitioned to milking in the AMS (Day 0), and on days 1, 2, 4, 8, 16, and 32 thereafter. As expected, primiparous cows (n=28) were significantly more likely than multiparous cows (n=49) to display step-kick behaviors both before and after teat attachment during milking ($P<0.05$). Elimination ($P<0.01$) and vocalization ($P<0.01$) were significantly greater on Day 0 compared to all other days. Milk yield significantly increased after Day 0 ($P<0.01$), and multiparous cows and cows earlier in lactation had significantly higher milk production on all days ($P<0.01$). The results suggest that multiparous cows may adapt more readily to milking in an AMS, potentially indicating the importance of age or milking experience. Significant increases in stress-related behaviors such as elimination and vocalization, as well as a significant decrease in milk yield indicate discomfort with the AMS on Day 0; however, the cows appear to adapt to the new milking system by Day 1.

Can a well-designed creep area reduce piglet mortality?

Vasdal, Guro[1], Glærum, Marit[1], Melišová, Michala[2], Bøe, Knut Egil[1], Broom, Donald M.[3] and Andersen, Inger Lise[1], [1]Norwegian University of Life Sciences, Animal and Aquacultural Sciences, Pb 5003, 1430 Ås, Norway, [2]2Department of Ethology, Institute of Animal Science, Uhříněves, 104 00 Prague, Czech Republic, [3]University of Cambridge, Centre for Animal Welfare and Anthrozoology, Department of Veterinary Medicine, Madingley Road, Cambridge CB3 0ES, United Kingdom; guro.vasdal@umb.no

High piglet mortality is still a concern in loose housed sows, and increased use of the creep area is commonly assumed to reduce piglet mortality. The aim of this study was to investigate, firstly, whether improving the thermal comfort and softness of the creep area would increase time spent in the creep area during the first three days after birth, and secondly, whether this would affect early postnatal piglet mortality in loose-housed sows. Forty-six loose-housed sows and their litters kept in individual farrowing pens were subjected to one of three creep area treatments during the first three days after birth (0-72h); CON; concrete floor in the creep area, BED; a thick layer (>10 cm) of sawdust on the creep area floor and HUT; a thick layer (>10 cm) of sawdust on the creep area floor in addition to an extra wall to increase the heat conserving capacity in the creep area. The sows were continuously video recorded from 1 day before farrowing until 3 days after farrowing. The behaviour and location of the piglets was scored using instantaneous sampling every 10 minutes from 08:00 - 14:00 and from 20:00 - 02:00 at day 0, day 1 and day 2. Piglets in HUT spent significantly less time in the creep area compared to piglets in CON and BED ($P<0.001$), while there was no difference in time spent in the creep area between CON and BED. Piglets in HUT spent significantly more time resting near the sow ($P<0.05$) than piglets in CON and BED. There was no significant difference in piglet mortality between the three treatments. Total time spent in the creep area had no significant relationship with piglet mortality. In conclusion, offering a heated creep area with soft bedding did not increase time spent away from the sow when the sow was not nursing, nor did it reduce piglet mortality. Quality of the creep area thus appears to have little impact on piglet mortality.

Effects of haul distance and stocking density on young suckling calves in Japan

Uetake, Katsuji[1], Tanaka, Toshio[1] and Sato, Shusuke[2], [1]Azabu University, School of Veterinary Medicine, Sagamihara, 229-8501, Japan, [2]Tohoku University, Graduate School of Agricultural Science, Osaki, 989-6711, Japan; uetake@azabu-u.ac.jp

Recommendations for calf transportation in the EU (distance less than 100 km) and the RSPCA (stocking density greater than 0.35 m^2/head) were evaluated their applicability to calves transported in Japan. Nineteen suckling calves (14 Holstein calves aged 11-26 days and five cross-bred (Japanese Black × Holstein) calves aged 13-45 days) were allocated to one of the stocking densities of 0.25, 0.35, 0.45 m^2/head and transported 50, 100 or 150 km at intervals of 7 days. Behavior and incidence of injury and watery feces were observed during and after transportation. Blood sampling was performed before and after transportation. The occurrence of calves laying and turning round during transportation was associated with stocking density (chi square test, both $P<0.05$), but not associated with haul distance. Total number of calves that laid down (8, 9, 14 calves at 0.25, 0.35, 0.45 m2/head, respectively) and the occurrence of turning round (0, 4, 6 times at 0.25, 0.35, 0.45 m^2/head, respectively) during transportation increased as it became less dense. Scratching on the knee was only observed in one calf. Incidence of watery feces after transportation was not associated with either haul distance or stocking density. There were significant effects of haul distance on the concentrations of plasma cortisol and noradrenaline, and serum AST and IgM (repeated-measures ANOVA, all $P<0.05$). These concentrations except cortisol were higher after transportation at 150 km compared to the pre-transportation value (Tukey's HSD test, AST and IgM $P<0.05$; noradrenaline $P<0.10$). Results suggest that haul distances greater than 100 km should not be recommended even for suckling calves transported in Japan.

Can we predict troubles during horse clinical examinations by a simple test?

Peeters, Marie, Godfroid, Sandra, Sulon, Joseph, Beckers, Jean-François, Serteyn, Didier and Vandenheede, Marc, University of Liège, Faculty of Veterinary Medicine, Boulevard de Colonster, 20, B43, 4000 Liège, Belgium; marie.peeters@ulg.ac.be

Hospitalization of horses involves important stress reactions that can cause injuries to medical staff and affect animal health and welfare. The aim of this study was to test the potentiality of a behavioural test (the weight-scale test WST) to predict stress reactions in horses during routine manipulations. Ten healthy horses, 7 mares and 3 geldings, from various breeds, aged from 3 to 26 (12.30±6.44) and weighing 498.90±84.78 kg were housed two days in the Equine Clinic of the University of Liege (Belgium) during spring 2009. On day 1, horses were filmed passing on a scale (black rubber flooring) and were observed during a hooves trimming (10'). On day 2, a vet performed a dental check (15'). Horses were scored for locomotion behaviours and body, head and feet movements. During WST, behavioural variables were scored (time, frequency and latency) using The Observer® (Noldus). During trimming and dental check, head and body movements were instantaneous sampled (every30'). Feet movements were continuously sampled. A 6-items questionnaire about horses' personality and exam quality, with a 5-Likert visual analogue scale, was fulfilled by the blacksmith and vet after manipulation. Significant Spearman correlations were found between behaviours observed during WST and behaviours observed during clinical examinations. Number of sudden head shaking during the WST are correlated to the number of sudden body moving during trimming (r_s,p) (0.79,0.004) and to the number of scraping the floor with the foot during dental check (0.67,0.022). The moving back time (%) during the WST is correlated to the number of scraping the floor during trimming (0.68,0.016). Number of trials needed to obtain the horse's weight is correlated with a poor handling quality (0.75,0.01) and a low education score (0.76,0.007) assessed by the blacksmith. Standardized, common and easy behavioural tests should thus be used to predict horses' handling quality in clinic.

Nursing behaviour in beef cattle: individual differences between cows, effect of parity, calf sex and birth weight

Stehulova, Ilona[1], Spinka, Marek[1] and Simecek, Petr[2], [1]Institute of Animal Science, Department of Ethology, Pratelstvi 815, 104 01 Prague, Czech Republic, [2]Institute of Animal Science, Biometric Unit, Pratelstvi 815, 104 01 Prague, Czech Republic; stehulova.ilona@vuzv.cz

The aim of the study was to investigate individual differences in nursing behaviour of beef cows as affected by parity, calf's sex and birth weight. The observations were carried out in 2001-2003 in a Gasconne cattle herd kept at the Institute of Animal Science in Prague. Nursing behaviour of 17 cows in 28 lactations (8 cows in one, 7 cows in two, 2 cows in three lactations) was directly observed for 12 hours at 3 and 30 days of calf's age. Number of nursings per 12 hours, mean nursing duration and the initiator and terminator of each nursing were recorded. Pearson correlations and a mixed linear model (SAS PROC MIXED) with a random effect of a cow were used to analyse the data. The fixed effects were day, year, parity, calf's sex and birth weight. We found individual differences between cows. They explained 42.8% of the variation in number of nursings and 40.2% in the cow's tendency to terminate nursing. Number of nursings (r=0.42, $P<0.05$) and proportion of nursings terminated by a cow (r=0.56, $P<0.01$) were correlated between days 3 and 30. Mean nursing duration and nursing initiation were not individually specific (<1.5% of the variation). The proportion of nursings initiated by the mother increased with parity ($P<0.01$) and the cows terminated nursing more often when the calf was a heifer than if it was a bull ($P<0.01$). There was significant interaction between parity and sex. High parity cows were nursing male calves more frequently ($P<0.05$) and terminated nursing less frequently ($P<0.01$) compared to low parity cows. Calf's birth weight did not influence cow's behaviour. In conclusion, there are individual differences in cows' nursing behaviour, which are constant between 3rd and 30th day after parturition. Cows on higher parities invest more in the calves, compared to low-parity ones and invest relatively more into male progeny.

The effects of cognitive experiments on the welfare of captive chimpanzees by behavioural measures

Yamanashi, Yumi[1,2], Hayashi, Misato[1] and Matsuzawa, Tetsuro[1], [1]Kyoto University, Section of Language and Intelligence, Kanrin, Inuyama, 4848506, Japan, [2]Japan Society for Promotion of Science, Chiyoda-ku, Tokyo, 1028471, Japan; yamanash@pri.kyoto-u.ac.jp

We investigated the effects of the cognitive experiments by direct comparison of activity budget between wild and captive chimpanzees. One goal of captive management is to ensure that the activity budgets of captive animals are as similar as possible to their wild counterparts. However, this has rarely been achieved. We compared the activity budget among three groups of chimpanzees; wild chimpanzees in Bossou (Guinea, N=10), captive chimpanzees who participated in the experiments (participant chimpanzees, N=6) and who did not participate in the experiments (non-participant chimpanzees, N=6) at Primate Research Institute (Japan). The cognitive experiments were conducted to test cognitive abilities of chimpanzees using touch panel screen. The chimpanzees manipulated the screen and small pieces of fruits were provided as rewards. The chimpanzees participated in the experiments spontaneously. The data from captivity were obtained on the experimental days (weekdays) and non-experimental days (weekends). In the both field sites, we followed each chimpanzee from about 7am to the time when chimpanzees started to rest in the evening. The behaviors were recorded every one minute. The total observation hours were 134.7hrs in Bossou and 227.8hrs in PRI. The results showed that on weekdays, feeding time and resting time of participant chimpanzees were almost the same with wild chimpanzees ($P>0.1$). However, as for non-participant chimpanzees, feeding time was significantly shorter and resting time was longer than those of wild chimpanzees ($P<0.05$). In contrast, no difference can be found in feeding time and resting time of the both groups of captive chimpanzees on weekends ($P>0.1$). The results suggested that the cognitive experiments work as an efficient tool for feeding enrichment. However, we observed abnormal behaviors (such as regurgitation and reingestion) during and outside of the experiments for participant chimpanzees. Based on the results, we will discuss the applicability and limitation of cognitive experiments for environmental enrichment.

The influence of environmental enrichment and personality on working and reference memory of pigs in a spatial discrimination task

Bartels, Andrea C., Oostindjer, Marije, Hoeks, Cindy W. F., De Haas, Elske N., Kemp, Bas and Bolhuis, J. Elizabeth, Wageningen University, Adaptation Physiology Group, Marijkeweg 40, 6709 PG Wageningen, Netherlands; fleur.bartels@wur.nl

Stimulus-poor rearing conditions of pigs may negatively affect their development and cognitive capacity. We assessed the impact of environmental enrichment on performance in a spatial discrimination task. Pigs were classified as high-resisters or low-resisters during a backtest, 1-min restraint in supine posistion. Pigs (n=32) were housed in stimulus-poor barren (B) or in larger, enriched (E) pens, with two high-resisters and two low-resisters, per pen. From 12 weeks they were exposed to a holeboard task, in which they had to discriminate four baited buckets out of 16 in a 4x4 matrix in a square arena with four entrances. Pigs were subjected to 30 trials with a maximum duration of 180 sec. The number of food-rewarded visits divided by total number of (re)visits to baited buckets was used to assess working memory (WM). Reference memory (RM) was calculated as the number of (re)visits to baited buckets divided by total number of visits. Block means of 5 trials were calculated and effects of housing and backtest classification were analyzed using GLMs, with values of piglets and pens in time-taken as repeated measurements for those effects. WM and RM improved over trials and trial duration decreased ($P<0.001$). B pigs visited the first baited bucket faster than E pigs (6.2 ± 0.3 vs. 7.1 ± 0.1 sec, $P<0.05$), but E pigs tended to have a shorter interval between visiting baited buckets ($P<0.10$). E pigs (0.90 ± 0.01) had a higher WM score than B pigs(0.87 ± 0.01) across trials($P<0.05$). Rearing environment did not affect RM. Low-resisters found on average more rewards, i.e. completed the task more often successfully ($P<0.05$) than high-resisters, particularly in the first 5-trial block (96.4 vs. 84.7%, $P<0.05$), but backtest classification did not affect memory scores. In conclusion, environmental enrichment did not affect reference memory, but improved working memory of pigs in a spatial discrimination task.

The effects of social and environmental enrichments on the welfare of tom turkeys
Weber, Patricia and Scheideler, Sheila, University of Nebraska-Lincoln, Animal Science, c206 AnSci 3800 Fair, 68583, USA; pweber2@unl.edu

The effects of broiler chick addition on the reduction of early mortality and the effects of providing environmental enrichment on the welfare of tom turkeys were determined. Exp. 1 consisted of 296 one-day-old turkey poults and 24 three-day-old broiler chicks. Sixteen pens were divided into 4 treatment groups (20 birds/pen). Trt 1 consisted of 20 poults/pen and were assisted in finding feed and water. Trts 2-4 were not assisted in finding feed or water. Trt 2 consisted of 20 poults/pen; Trt 3 consisted of 18 poults and 2 chicks/pen and trt 4 consisted of 16 poults and 4 chicks/pen. Body wt., feed intake, mortality and time budget data (average daily % time spent--eating, drinking, active, resting) were collected. Exp. 2 utilized 288 12 day-old turkeys divided into eight pens (36 birds/pen). Four pens were enriched with a 1.5m^2 platform with side rails and a ramp. Two perches were provided, one 73cm above the platform and one 55cm above the floor. The remaining 4 pens were barren. Body wts. feed intake, gait scores, mortality, time budget and motivational data (% time using enrichments) were collected. No significant treatment differences were observed for feed intake, body wt. or mortality in exp. 1 or 2. Average daily % time being active was significantly increased with the addition of 4 broiler chicks (trt 4) when compared to assisted poults (trt 1), 40.4% and 46.1%, respectfully ($P<0.07$). In exp. 2, no significant differences were observed in time budgets between treatments. At 5 weeks of age 72% of the birds visited the enrichments during the day and at 10 weeks 39% visited the enrichments. No productivity parameters were improved or declined with the addition of the environmental enrichments; however a strong motivation to climb, perch and play was observed.

Evaluation of identification methods for mice

Spangenberg, Elin[1,2], [1]Swedish University of Agricultural Sciences, Department of Animal Environment and Health, P.O. Box 234, 532 23 Skara, Sweden, [2]Karolinska Institutet, Veterinary Resources, 171 77 Stockholm, Sweden; elin.spangenberg@ki.se

Individual identification of animals is important to ensure valid and reliable results in scientific studies. Identification (ID) of mice is considered routine management and few studies have focused on animal welfare aspects. This study evaluated acute effects of the ID methods partial toe amputation (TOE), ear tagging (TAG) and ear punching (PUNCH) in preweaning mice. Ten litters of the FVB/N strain were used (83 pups). Two pups/litter were assigned to each method and two pups/litter were controls. The morphological, motor and sensory development of the pups was monitored from birth (day1) to weaning (day21) using an early characterization protocol. TOE was performed day7 or 8 by cutting one toe at the distal part of the second phalangeusing sharp scissors. TAG and PUNCH were performed day19, by attaching a metal ear tag to, the pinna or punching a hole on the edge of the pinna, respectively. TAG animals showed the strongest reaction; 62% vocalised during the procedure and frequent behaviours were shaking and jumping. The pinna was red and swollen for three days. In the PUNCH group 19% vocalised during the procedure and behaviours observed were scratching the ear (33%) and freezing (14%). In the TOE group 29% vocalised during the procedure. All pups crawled into the nest in the cage and one pup licked its toe. The toes were swollen for three days. Morphological, motor and sensory development was the same for all treatment groups. On day21, a clinical examination was performed on all pups. The main findings were; all groups were scored as normal for gait, grip strength and balance. However, all TAG pups had abnormal body posture with tilted heads, and the highest prevalence of provoked biting. In summary, TAG resulted in clinical and behavioural signs of pain and discomfort and is therefore the least preferred method of those tested here.

Maternal hormones and avian primary offspring sex ratio: a review

Goerlich, Vivian[1] and Groothuis, Ton[2], [1]Linköping University, Applied Ethology, Linköping University, 58183 Linköping, Sweden, [2]University of Groningen, Behavioural Biology, Biological Center, Kerklaan 30, 9751 NN Haren, Netherlands; vivgo@ifm.liu.se

Deviations in offspring sex ratio (the proportion of sons and daughters) from the expected 50% have been frequently documented in several vertebrate species and the ultimate and proximate causes have fascinated researchers for centuries. Parental physiology and behaviour is shaped by the surrounding environment which often correlates with offspring sex ratio biases. However, the underlying mechanisms remain largely elusive. Apart from understanding the evolutionary causes, identification of the proximate mechanisms would greatly benefit applied sciences such as conservation biology and the commercial breeding industry. In birds, females are the heterogametic sex (ZW) while males provide sperm holding only the Z chromosome. This enables the mother to influence the sex of the ovum even prior to actual fertilisation. Given their sensitivity to environmental cues, their involvement in the reproductive physiology, and the substantial amounts present in the developing ovum, maternal steroid hormones (e.g., testosterone and corticosterone) are prime candidates for mediating primary sex ratio adjustment. In this review we first introduce meiotic drive (non-random chromosome segregation) as the prime candidate mechanism for primary sex ratio manipulation. We then discuss the current evidence for the involvement of maternal hormones, incorporating findings from our recent studies with those of other correlative and experimental studies. We manipulated maternal plasma hormones in homing pigeons (*Columba livia*) and found, in line with evolutionary predictions, that elevated testosterone seems to induce a male biased and corticosterone a female biased primary sex ratio. As yet we found no strong evidence for a role of maternal hormones in the egg yolk. However, several results of the reviewed studies are inconsistent with each other; therefore we briefly discuss methodological issues that may explain these discrepancies. Finally, we point out suggestions for future research involving *in vitro* studies and the use of molecular and genetic techniques.

Using stimulus-induced behavioural tests to assess stereotypies and abnormal behaviours in sows

Devillers, Nicolas[1], Kerivel, Anna[1], Gète, Maud[1], Bergeron, Renée[2] and Meunier-Salaün, Marie-Christine[3], [1]Agriculture and Agri-Food Canada, Dairy and Swine R&D Centre, P.O. Box 90, J1M 1Z3 Sherbrooke, QC, Canada, [2]University of Guelph, Alfred Campus, P.O. Box 580, K0B 1A0 Alfred ON, Canada, [3]INRA, SENAH, Domaine de la Prise, 35590 St-Gilles, France; nicolas.devillers@agr.gc.ca

Stereotypies are usually interpreted as abnormal behaviours and a sign of poor welfare. The aim of this study was to develop an easy-to-use procedure for on-farm assessment of abnormal behaviours in gestating sows. Behaviour of 51 individually housed sows was observed after the morning feed supply during ten 3-min focal observations. Sows were thereafter categorized as either stereotypers when they expressed repetitive sequences of behaviour (ST, n=36), or sham-chewers (SC, n=33) or chain-chewers (CC, n=26) when they spent more than 15 or 5% of time sham-chewing or chain-chewing, respectively. In the afternoon, each sow was submitted to three tests consisting of giving and removing a stimulus: 20 g of feed (Feed), 1000 cm^3 of straw (Straw) or 300 cm^2 of rubber mat (Mat), applied as a frustrating situation to elicit abnormal behaviours. Behaviour was continuously observed before, during and after application of each stimulus. The Glimmix procedure of SAS was used to compare sows belonging to a category with those that did not. NoST (n=15) and NoSC (n=18) sows showed an increase in chain-chewing after Mat removal while ST and SC sows did not ($P<0.05$). CC sows always spent more time chain-chewing independently of the stimulus ($P<0.01$). They spent also less time sniffing the floor ($P<0.01$) or licking the floor or the feeder ($P<0.05$) after Straw removal than NoCC sows (n=25). Increase in floor-sniffing was greater in SC sows after Straw removal ($P<0.05$). In summary, removal of Mat induced chain-chewing but not in ST or SC sows. Similarly, removal of Straw induced licking and floor-sniffing but less in CC sows and more in SC sows. These results show that patterns of chronic stereotypies may modulate response of sows to behavioural challenges. This approach of inducing behaviour with a stimulus could be a practical method of assessing sow stereotyped behaviours.

Aggression toward people by Rottweiler dogs

Heinonen, Sanni[1], Malkamäki, Sanna[1] and Valros, Anna[2], [1]Research Center of Animal Welfare, Department of Equine and Small Animal Medicine, Koetilantie 7, P.O. Box 57, 00014 University of Helsinki, Finland, [2]Research Center of Animal Welfare, Department of Production Animal Medicine, Koetilantie 7, P.O. Box 57, 00014 University of Helsinki, Finland; sanni.heinonen@helsinki.fi

Several serious aggressive encounters towards people by rottweiler breed dogs have been reported in recent years in Finland. To investigate mechanisms behind this, we studied personality traits and types of aggression of rottweilers with an online questionnaire directed to dog owners. The questionnaire consisted of 110 questions of which 102 focused on the dogs´ behavior in everyday situations and 8 questions on the owners´ impression of the temperament of the dog. All questions were answered on a five-point scale (1 = no reaction, 5 = strong or frequent reaction). 544 replies were received, of which 514 dogs more than 2 years old were selected to the final analysis. The data was analyzed by principal component analysis with promax rotation which allows for correlation between the components. Correlations between different components were calculated with Spearman's correlation. We extracted 14 different components (Eigenvalue > 1,5), including aggression towards strange people, fear of strange people, aggression in handling situations, aggression against strange dogs, excitability, social aggression towards people, pain sensitivity, playfulness, aggression towards owner, noise sensitivity, separation anxiety, emotional reactivity, territorial aggression and attention seeking behavior. Aggression towards strange people correlated significantly with several other components, such as with territorial aggression (r=0,587, $P<0.01$), fear of strange people (r=0,291, $P<0.01$), aggression in handling situations (r=0,337, $P<0.01$), social aggression toward people (r=0,267, $P<0.01$) and aggression against strange dogs (r=0,411, $P<0.01$). All aggression related components, except on territorial aggression, correlated with the fear-related components or included fear related items. We concluded that aggression toward people by rottweiler dogs is strongly motivated by fear and defensive characteristic of the breed.

The effect of anaesthesia and analgesia on the behaviour of piglets after castration

Sutherland, Mhairi[1] and Davis, Brittany[2], [1]AgResearch, Animal behaviour and welfare, Ruakura Research Centre, Hamilton 3123, New Zealand, [2]Texas Tech University, Animal and Food Sciences, Animal and Food Sciences Bldg, Lubbock, 79409, USA; mhairi.sutherland@agresearch.co.nz

Surgical castration is routinely conducted on commercial swine farms to prevent boar taint and reduce aggressive behaviours. However, the procedure of castration causes acute pain which is an animal welfare concern. The objective of this study was to evaluate the combined effect of anaesthesia and a non-steroidal anti-inflammatory drug on the pain-induced distress caused by castration in piglets. At 3 days of age piglets were allocated to one of five treatments: (1) Sham castration (CON; n=5); (2) Surgical castration (CAS; n=5); (3) Surgical castration while the pig was anaesthetized with carbon dioxide gas (G+CAS; n=5); (4) Surgical castration plus NSAID administered at the time of castration (N+CAS; n=5); and (5) Surgical castration conducted while the pig was anesthetized with carbon dioxide plus NSAID administered at the time of castration (G+N+CAS; n = 5). Piglet behaviour was recorded (live observations) in the farrowing crates using 1-min scan sampling for up to 120 min after castration. Data were analyzed using the MIXED procedures of SAS. During the first 15 min after castration, all castrated piglets spent more ($P<0.05$) time performing active behaviors compared with CON piglets (CON: 0.31±0.04; CAS: 0.45±0.04; G+CAS: 0.46±0.04; N+CAS: 0.44±0.04; G+N+CAS: 0.39±0.04). After 2 h, CAS piglets spent more ($P<0.05$) performing active behaviours compared with CON piglets and piglets given anaesthesia, analgesia, or both prior to castration displayed similar ($P>0.05$) levels of activity as CON piglets (CON: 0.01±0.04; CAS: 0.09±0.04; G+CAS: 0.03±0.04; N+CAS: 0.01±0.04; G+N+CAS: 0.03±0.04). Anaesthesia and/or analgesia did not appear to influence the initial behaviour of piglets to castration, but may have had some beneficial effect on the behaviour of piglets by 2 h after castration. Further research is needed before any conclusions can be made on the effectiveness of these analgesic treatments in reducing the pain-induced distresses caused by castration.

CowLog: a free software for analyzing behavior from video

Pastell, Matti[1,2] and Hänninen, Laura[2,3], [1]University of Helsinki, Department of Agricultural Sciences, P.O. Box 28, 00014 University of Helsinki, Finland, [2]University of Helsinki, Research Center for Animal Welfare, P.O. Box 57, 00014 University of Helsinki, Finland, [3]University of Helsinki, Department of Production Animal Medicine, P.O. Box 57, 00014 University of Helsinki, Finland; matti.pastell@helsinki.fi

Most behavior observations rely on video recordings because direct observation has a high risk of disturbing the observed subject. The observer scores the video data, usually with the help of behavior coding software, which are usually costly. We have developed CowLog, which is open-source software for recording behaviors from digital video. CowLog is coded in Python programming language and it uses the PyQt toolkit for the graphical user interface and MPlayer media player for playing the digital video. It works on all major operating systems including Windows, Linux and Mac OS X and supports all the common video formats as well as highspeed video. It can be freely downloaded from its website at http://mpastell.com/cowlog. The program has two main windows: a coding window, which is a graphical user interface used for choosing video files and defining output files that also has buttons for scoring behaviors, and a video window, which displays the video used for coding. The windows can also be used in separate displays. The user types the key codes for the predefined behavioral categories, and CowLog transcribes their timing from the video time code to a data file. The software supports using up to 24 state or event behaviors. No initial setup is needed, since the key code definitions are always the same; however, the user needs to keep track of codes elsewhere. Coding is performed continuously by logging the time each behavior starts. In the analysis stage, the type of the behavior is given as an input for the analysis function, and the start time of the next state is used as the end time of the previous state. Accordingly, the start time of events is used to mark the occurrence of a behavior. CowLog comes with an additional feature, an R package called Animal, for elementary analyses of the data files. With the analysis package, the user can calculate the frequencies, bout durations, and total durations of the coded behaviors and produce summary plots from the data.

The effect of topical anaesthesia on pain alleviation in calves post-dehorning

Espinoza, Crystal, Lomax, Sabrina and Windsor, Peter, The University of Sydney, Faculty of Veterinary Science, 445 Werombi Rd, Camden NSW 2570, Australia; crystal.espinoza@sydney.edu.au

Two trials were conducted to examine the effects of topical anaesthesia on the pain response of heifer dairy calves experiencing dehorning. Trial 1 involved the observation of pain-related behaviour following treatment with or without topical anaesthetic (Tri-Solfen®, Bayer Animal Health, Australia) following hot-iron dehorning, or sham handling control. Behaviour was documented by an observer blind to treatment from 30 min to 5 hours post-dehorning. A Numerical Rating Scale (NRS) was used to categorise behaviour into nil, mild, moderate and severe (0, 1, 2 and 3 respectively) displays of pain-related behaviour. Calves hot-iron dehorned and given topical anaesthetic displayed lower ($P=0.023$) NRS scores than calves dehorned without anaesthetic. There was no difference ($P=0.277$) in NRS scores between calves hot-iron dehorned and given topical anaesthetic, and control calves. Trial 2 used Wound Sensitivity Testing (WST) to determine the sensitivity of the dehorned wound surface following scoop dehorning with or without the administration of topical anaesthetic (Tri-Solfen®). Von Frey monofilaments, calibrated at 10 and 300 g, were used to provide light touch and pain stimulation respectively, to the wound and peri-wound area before, and up to 3 hours post-dehorning. A NRS was used to grade responses depending on vigour. Calves treated with topical anaesthetic displayed no difference ($P=0.134$) in response to light touch stimulation of the wound compared to untreated calves. There were tendencies for treated calves to exhibit lower scores when stimulated at the peri-wound site by light touch ($P=0.06$) and pain ($P=0.051$) stimulation. In response to pain stimulation of the wound, treated calves displayed lower scores than untreated calves ($P=0.01$). The results from these trials suggest the use of topical anaesthesia has the potential to reduce the pain-related behaviour associated with hot-iron dehorning and to reduce the sensitivity of the dehorned wound site post scoop dehorning in calves. This novel method of pain relief can provide a more practical and affordable alternative to options currently available.

Hay consumption by pigs of Krskopolje breed and modern hybrid

Zupan, Manja, Žemva, Marjeta, Malovrh, Špela and Kovač, Milena, University of Ljubljana, Biotechnical Faculty, Department of Animal Science, Groblje 3, 1230 Domžale, Slovenia; manja.zupan@bfro.uni-lj.si

Krskopolje breed is the endemic pig breed in Slovenia and is usually reared with low financial input. This study is the first to investigate the feeding behaviour of Krskopolje pigs. Hay was given daily to Krskopolje pigs and pigs of a modern hybrid (1122) as rooting material and nutritional substrate. The aim was to find out if there are differences in hay consumption between the two genotypes. Fattening pigs (N=24) of both sexes and genotypes were reared on straw bedding in 8 pens, each containing 3 animals of the same sex and genotype. Pigs were fed once daily by hand with commercial feed and had free access to hay. The amount of feed was calculated to meet NRC (1998) requirements of growing-finishing pigs. Pigs were observed directly on four consecutive days from 12h to 16h using continuous recording in 5 min periods. All pigs consumed hay, most often while standing, and no differences in the positions were observed between the genotypes ($P>0.10$, LR test). The modern hybrid pigs consumed hay more frequently (% of time) than pigs of the Krskopolje breed (47.9 vs. 29.9, respectively; $P<0.001$; LR test) regardless of sex (37.5 vs. 41.2; n.s.) or day (38.9 vs 41.7% vs. 38.9 vs. 36.1; n.s.). Their consumption was most frequent during the first observational period, i.e. from 12:00 to 12:40 (79.2% of time). However, Krskopolje pigs consumed hay most frequently during the fourth observational period, i.e. from 14:00 to 14:40 (43.8%; $P<0.001$). Pigs of the modern hybrid tended to consume hay more synchronously than pigs of the Krskopolje breed (averaged nr. of pigs: 1.9 vs. 1.5, respectively; $P<0.10$; LR test). The results show that hay consumption varied between pigs of the two genotypes, possibly reflecting a different experience with competition at feeding.

Factors influencing nociceptive pressure thresholds in piglets

Janczak, Andrew Michael[1], Hild, Sophie[1], Fosse, Torunn Krangnes[1], Ranheim, Birgit[1], Andersen, Inger Lise[2] and Zanella, Adroaldo J.[1], [1]The Norwegian School of Veterinary Science, P.O. Box 8146, Dep N-0033, Oslo, Norway, [2]Norwegian University of Life Sciences, P.O. Box 5003, No-1432, Ås, Norway; andrew.janczak@nvh.no

Measurements of nociceptive pressure thresholds can be used both for testing the efficacy of anaesthetics and analgesics, and for assessing hyperalgesia in chronic pain states in research and clinical settings. The present experiment aimed at describing the repeatability of these measurements and how different factors contribute to their variability. Forty-four piglets from four different pens were tested for pressure thresholds twice on each of four different days distributed over the first and second week of life. Piglets were retrieved from their home pens in random order and placed in a sling. A Commander™ Algometer tip (the tip of a device used for measuring pressure thresholds; JTECH Medical) was applied to the metacarpus/metatarsus of the piglets' legs to measure the pressure at which piglets retracted their leg, at which time stimulation was terminated, with a safety cut off of 25 Newtons. We tested the mixed model: threshold = pen + week + day(week) + repetition(day) + piglet weight + behaviour prior to testing. Sensitivity increased over repetition within days ($P<0.001$), and over days within each week ($P<0.002$), but decreased over weeks ($P<0.0001$). Heavier piglets were less sensitive ($P<0.0001$), possibly due to changes in the mechanical properties of tissue, and activity prior to testing was associated with reduced sensitivity ($P<0.05$). The Pearson correlation between repeated measures within days 1-4 was 0.47, 0.60, 0.56, and 0.78, respectively. The correlation between the first measure on day 1 and day 4 was 0.43. The results indicate that a range of factors (weight, behaviour prior to testing, and experience) need standardization in order to reduce variability in studies using nociceptive pressure thresholds in piglets. It is also clear that measures taken in naïve piglets are not highly predictive of measures taken in more experienced piglets. The repeatability of measures did, however, increase to an acceptable level with repeated exposure to the testing protocol, suggesting that this methodology should be useful for testing the efficacy of anaesthetics and analgesics.

Quantifying hungry broiler breeder dietary preferences using a closed economy T-maze task
Buckley, L.A., Sandilands, V., Tolkamp, B.J. and D'Eath, R.B., Scottish Agricultural College, West mains Road, Edinburgh., EH9 3JG, United Kingdom; louise.buckley@sac.ac.uk

This study aimed to identify hungry broiler breeders' (n = 12) preferences for quantitative (Control) or qualitative (QDR) dietary restriction in a closed economy environment. The QDR option was the control diet with the addition of either 3 g Calcium Propionate/kg total feed (CVC treatment group) or 300 g oat hulls / kg total feed (CVF treatment group) (both: n=6). Control and QDR portions ensured equal growth, regardless of choice. Each bird was initially trained with a Control diet versus no food task and a QDR diet versus no food task to allow birds to learn the satiating properties of each diet. Birds had to associate the T-maze coloured arms with dietary outcomes to immediately obtain food. Birds learnt this task easily (CVC: $\chi2$ = 59.7, P<0.001; CVF $\chi2$ = 0.59.7, P<0.001). A choice between the Control diet and the QDR diet was then offered but neither treatment group demonstrated a diet preference (CVC: $\chi2$ = 0.35, P>0.1; CVF $\chi2$ = 0.23, P>0.1). Study modifications demonstrated this was not a failure to discriminate between the diets per se (mean control diet / total diet consumed over 30 minutes = 117g/142g, P<0.001) or novel colour combination confusion (CVC: $\chi2$ = 36.2, P<0.001; CVF $\chi2$ = 46.49, P<0.001). Most (seven out of twelve) birds still failed to show a significant preference when the Control diet quantity was increased by 50% to make it 'obviously' bigger and more desirable (CVC: $\chi2$ = 0.01, P>0.1; CVF $\chi2$ = 2.25, P>0.1). Therefore, it was concluded that the failure to show a dietary preference was due to task learning failure and not lack of dietary preference. Possible reasons for this failure are discussed.

Using participatory approaches in applied animal welfare research: examples from working equine welfare

Eager, Rachel, A. and Childs, Amanda, C., The Brooke, 30 Farringdon Street, EC4A 4HH London, United Kingdom; rachel@thebrooke.org

Applied ethology research, when used to inform policy and practice, can improve the welfare of the animal as the intended beneficiary. It's applicability and impact can be greatly improved, however, when the role of the owner/farmer (stakeholder) is also considered. Over the last 20 years, development workers and researchers have begun to recognise the importance of stakeholder participation in development and research. 'Participation' describes the level of involvement/control stakeholders have over the research process, ranging from being entirely researcher-led (contract research) to stakeholder-led (collegiate research). This presentation will discuss the use of participatory approaches to improve research in the context of analysing risk factors in working equine welfare. The Brooke is a working equine welfare charity aiming to develop evidence-based interventions on priority welfare issues. Field-based survey techniques are used to identify potential causes of poor welfare and provide evidence to inform interventions by community engagement and veterinary staff. Research examines environmental, resource, animal and human-based measures which may be directly or indirectly associated with the welfare issue. Participatory methodologies are incorporated throughout the research process. Examples include: (1) Matrix ranking: understanding owners prioritization of welfare issues (2) Social mapping: identifying animal owning households and informing sampling (3) Focus group discussions: improving contextual understanding (guiding data collection strategies); building trust and understanding with owners (4) Root cause analysis: identifying previously unknown risk factors (5) Activity mapping: understanding working patterns and daily lives of owners and equines. Researchers consider the incorporation of participatory approaches to have improved the quality of the research process by enabling increased contextual understanding; incorporation of community perspectives of risk; facilitating data collection within the community and increasing the applicability of findings to owners and field staff. Ultimately, research findings are more likely to be accepted by owners and used to improve the welfare of their animals.

Topical anaesthesia for the amelioration of mulesing pain in sheep

Lomax, Sabrina[1], Windsor, Peter[1] and Sheil, Meredith[2], [1]University of Sydney, Veterinary Science-Farm Animal Health, C02- Camden Buildings, University of Sydney, NSW 2006, Australia, [2]Animal Ethics Pty Ltd., P.O. Box 363, Yarra Glen, VIC 3775, Australia; sabrina.lomax@sydney.edu.au

Mulesing, the practice of cutting loose folds of skin from the breech area of sheep, is a common husbandry procedure routinely performed on Merino lambs annually in Australia. This has been proven to cause acute pain and stress, which has led to a growing concern for the welfare of lambs undergoing this procedure. We have undertaken studies that have shown that topical anaesthesia (TA) can significantly reduce wound pain, and improve wound healing and recovery in the first 8 hours and have furthered our studies to include up to 24 hours post-mulesing. Forty-two mixed sex Merino lambs (21.04 ± 0.5 kg) were randomly allocated to one of three treatment groups for either un-mulesed control (n=14) mulesing only (n=14), or mulesing in combination with TA (n=14). At the conclusion of the trial, all mulesed lambs were treated with TA to ensure improved welfare. Time for lambs to mother-up and feed, pain-related behaviours and wound pain (using von Frey monofilaments and numerical rating scale) were assessed in the 24-hour period post-mulesing. Results were analysed using analysis of variance for repeated measures (REML) and linear regression. Time to mother or feed didn't differ significantly between treatment groups ($P>0.05$). Pain behaviours and wound pain were assessed using a customized numerical rating scale (NRS). TA treated lambs displayed significantly lower pain-related behaviour scores compared with untreated lambs at both 1 hour (0.2 vs. 1.2, $P=0.002$) and 24 hours (0 vs. 1, $P=0.003$) post-mulesing. Response scores to pain stimulation of the wound surface were significantly lower in Tri-Solfen® treated lambs (2.44) and unmulesed lambs (2.2) than untreated lambs (14.8), $P<0.001$. These results indicate the use of topical anaesthesia as a cost-effective and management friendly tool for improving the welfare of livestock undergoing painful husbandry procedures.

The effects of Body Condition Score on stress and metabolic responses to a cold challenge in pregnant ewes: preliminary results

Verbeek, Else[1,2], Waas, Joseph[2], Oliver, Mark, H.[3], McLeay, Lance, R.[2] and Matthews, Lindsay, R.[1], [1]AgResearch Limited, Animal Behaviour and Welfare, East Street, Private Bag 3123, Hamilton, New Zealand, [2]University of Waikato, Department of Biological Sciences, Hillcrest Road, Private Bag 3105, Hamilton, New Zealand, [3]University of Auckland, The Liggins Institute, Private Bag 92019, Grafton, Auckland, New Zealand; else.verbeek@agresearch.co.nz

The health and welfare of animals is dependent on their ability to adapt to changes in the environment in order to maintain energy homeostasis. Grazing sheep are likely to be subjected to environmental challenges such as food restriction and cold. We aimed to assess stress and metabolic responses in 18 shorn ewes with three different body condition scores (BCS: low, LBC; medium, MBC; high, HBC) to a six hour acute cold challenge in mid-gestation. During the cold challenge, animals were placed in a climate controlled room (5 °C) equipped with water sprinklers and wind fans. Blood samples were collected via jugular catheters. Data were analysed by ANOVA and the height of the peak responses, the rate at which these were reached and the area under the curve (AUC) were tested. Cortisol peak height and rate were reduced in LBC ewes (17.1±0.95 ng/ml and 0.28±0.05 ng/ml min^{-1}, respectively) compared to MBC (34.0±4.86 ng/ml and 0.90±0.11 ng/ml min^{-1}, respectively) and HBC (31.6±3.76 ng/ml and 0.77±0.12 ng/ml min^{-1}, respectively) ewes ($P=0.015$ and $P=0.006$, respectively). There was a tendency for an effect of BCS on AUC for cortisol ($P=0.079$). Insulin peak was lower in LBC ewes (0.18±0.02 ng/ml) compared to MBC (0.30±0.12 ng/ml) and HBC (0.30±0.05 ng/ml) ewes ($P=0.012$). LBC (0.007±0.002 ng/ml min^{-1}) and MBC (0.008±0.002 ng/ml min^{-1}) ewes also took longer to mobilize free fatty acids (FFA) than HBC (0.018±0.003 ng/ml min^{-1}) ewes ($P=0.009$). BCS also significantly affected AUC for FFA ($P=0.014$). The low insulin response in LBC ewes allowed circulating glucose levels to be maintained and supported energy homeostasis. In conclusion, the BCS of pregnant ewes significantly affected stress and integrated metabolic responses to a cold challenge which may have implications for the welfare of ewes with a low BCS.

Animal welfare risk assessment: stating the problem

Maijala, Riitta, Director of Risk Assessment, European Food Safety Authority, Largo N. Palli 5/a, 43100 Parma, Italy; riitta.maijala@efsa.europa.eu

Harmonised EU rules are in place covering a range of animal species and welfare-related issues. Council Directive 98/58/EC lays down minimum standards for the protection of all farmed animals, while other EU legislation sets welfare standards for different species and stages of animal production. The overall framework for EU action is set out in rolling action plans, currently the Community Action Plan on the Protection and Welfare of Animals 2006-2010. The European Commission (EC) has also adopted a Communication setting out the EU's Animal Health Strategy (AHS) for 2007-2013. The proposals include the development of an EU food label for animal welfare. As the International Committee of the World Organisation for Animal Health (OIE) has recognised, animal welfare is a complex, multi-faceted public policy issue which includes scientific, ethical, economic and political dimensions. This nature of animal welfare and the situation, where many of the welfare problems among animals are caused by human activities, make it difficult to separate purely scientific facts from the perceptions and ethical values of individuals. Applying the risk analysis framework can contribute to solving this problem. EFSA is in the forefront for developing risk assessment methodologies for animal welfare risk assessment. EFSA's activities in this area are carried out by the Panel on animal health and welfare (AHAW). Its scientific opinions focus on helping risk managers to identify methods to reduce unnecessary pain, distress and suffering for animals and to increase animal welfare where possible. This work will support the definition of an acceptable level of risks by the society. This is especially challenging since animal welfare is an area where views tend to be diversified, i.e. there is no single acceptable level of risk for an entire country and even less so for the whole European Union.

Expert elicitation in animal welfare risk assessment: methods and limitations

Bracke, Marc B.M., Animal Welfare Scientist, Livestock Research, Wageningen University and Research Centre, P.O. Box 65, 8200 AB, Lelystad, Netherlands; marc.bracke@wur.nl

The objective of both risk assessment and semantic modelling of animal welfare is to support political and ethical decision making. In semantic modelling procedures have been designed to systematically transfer scientific knowledge into welfare scores. Principles developed for semantic modelling may benefit risk assessment, e.g. using expert elicitation of welfare scores and weighting factors to 'validate' risk assessment outcomes. Expert elicitation can help to increase transparency as experts may have diverging views on welfare, as was found for veterinary and behavioural experts assessing housing systems for calves. In addition, for policy makers and the general public it is difficult, if not impossible, to comprehend how expert opinions as formulated in ESFA risk assessment reports will translate into improved welfare. The case of tail biting and tail docking in pigs is used as an example to show that expert opinion is crucial to further translate the EFSA report into risk management decisions, be it as a tool to solve welfare problems on farms or as an update of EC Directive 2001/93 on pig welfare. While the Directive may be revised radically from existing environment-based prescriptions into exclusive use of animal-based welfare measures such as curly pig tails, as suggested by the Welfare Quality® project, it may be worthwhile to consider alternatives perhaps involving only moderate adjustments of existing regulations together with enhanced compliance in order to bring about the required improvements in pig welfare.

Animal welfare risk assessment in national official control: a practical approach

Hultgren, Jan and Algers, Bo, Department of Animal Environment and Health, Swedish University of Agricultural Sciences, P.O. Box 234, 532 23 Skara, Sweden; jan.hultgren@hmh.slu.se

Because the European Food Safety Authority (EFSA) should 'promote and coordinate the development of uniform risk assessment methodologies in the fields falling within its mission', EFSA has in its reports on animal welfare issues produced Animal Welfare Risk Assessments (AWRAs). Scientific experts have compiled comprehensive lists of hazards and negative consequences, generated extensive reports on the state-of-art and made these readily available to everyone. However valuable this information is to policymakers and stakeholders, these exercises, which constitute tremendous efforts, have not brought about a deeper understanding of the consequences of alternative actions for the overall welfare status of farm animals. Following a self-mandate in 2007, EFSA is trying to further develop guidelines on risk assessment of animal welfare. One obstacle, though, is the lack of a common definition of good animal welfare. Several optional definitions exist, and the choice of definition is bound to be based on personal values; although conflicting theoretical views might often result in similar judgements in practice. Another obstacle is confusion regarding the animal welfare endpoints to consider. Some factors in animal environments could be regarded both as hazards and as indicators of poor welfare. The choice and characterization of animal welfare hazards is therefore also partly arbitrary. A third obstacle is the assumed existence of multiple interactions between different hazards, which are virtually impossible to handle mathematically and largely have to be ignored. These obstacles are to some extent possible to overcome by taking a different approach. A methodology is presented to collect information from experts, stakeholders and authority representatives on the assessed risk of poor animal welfare in different types of Swedish animal husbandry, making it possible to compare different husbandry categories, different risk questions and different assessors, to calculate overall estimates and to risk classify husbandry categories as a basis for official control.

Technical prospects for automated behavioural monitoring

Pastell, Matti, Department of Agricultural Sciences, P.O Box 28, 00014 University of Helsinki, Finland; matti.pastell@helsinki.fi

Precision Livestock Farming is a field of science that applies technological methods to monitor and control animals and their environment at an individual level. Recent developments in electronics, sensor design and computation have also inspired a lot of new interesting applications in automated behaviour and health monitoring of animals. Several sensors have been applied for monitoring animal behaviour. Accelerometers, Radio frequency identification (RFID), image and sound analysis are some of the potential techniques that with appropriate models can be used to predict what the animal is doing. Behaviours such as lying, standing and eating can already be measured from many different species in a reliable way, but more subtle changes, such as resting head postures, tongue rolling or social behaviour, are still difficult to detect. Current commercial and research systems are usually based on rather small number of animals and a specific environment, because of the high cost associated to data collection. Depending on the measurement this can cause problems when they are introduced to a different environment (e.g. with different breed, barn type, climate) and the produced results might not be reliable. This is a common problem with data based models also in other fields of science that is inherent with the modelling process. Therefore the validation of an automatic tool in the current research setting is highly advisable. Automated methods for monitoring animals show promise in better detection of health problems in practical farming conditions and bring a new useful tool for animal welfare research. However, special care should be taken that the used methods are scientifically validated.

Automated farrowing prediction and birth surveillance in farm animals

Pedersen, Lene J., Malmkvist, J. and Jørgensen, E., Århus University, Faculty of Agricultural Sciences, Denmark, Blichers Allé 20, DK-8830 Tjele, Denmark; Lene.juulpedersen@agrsci.dk

Neonatal mortality is high in many farm animal species. The average mortality rate in DK is now 23% for piglets, 6% for calves and around 10-20% for mink kit. Not only does the survival rate of the young to a great extent determine the farmer's economics, but high mortality rates are also of great concern in relation to animal welfare. In sow herds it has been shown that birth surveillance can reduce mortality significantly, both through securing piglets from hypoxia due to prolonged birth intervals and through avoiding hypothermia and low colostrum uptake in newborn piglets. However, since birth surveillance is time consuming and labour cost is high these beneficial interventions are not carried out in larger herds. Thus, there is a great need for technological advances that can help the farmer guide his/her attention towards problem animals. Currently we are developing methods to predict the time of farrowing *before* it actually occurs. The prediction is based on automated sampling of nesting behaviour by sensors. When nesting behaviour is detected a signal will be send to the climate control system that automatically switch the supplementary heating to the new born piglets on and off based on the farrowing status of the individual sow. To be used *during* birth, sensors (image analysis) are developed that can detect the birth of each individual piglet. By linking it to an alarm system the farmer can be warned in cases of birth problems in a specific pen. In the period *after* birth detection of weak piglets based on their lying position (image analysis) and/or surface temperature will be developed as well as methods to detect sows with heath problems based on their behavioural patterns. These methods can assure a timely intervention for animals in need and an undisturbed natural parturition for those animals that are not in need.

Automated detection of lameness in cows

Rushen, Jeff[1], Chapinal, Nuria[2] and De Passillé, Anne Marie[1], [1]Agriculture and Agri-Food Canada, Agassiz, BC, Canada, [2]University of British Columbia, BC, Canada; jeff.rushen@agr.gc.ca

Lameness in dairy cows is a major health and welfare problem but farmers have difficulty detecting lame cows. Because farms are getting larger with fewer workers there is a need for automated methods of detecting lameness. Lameness in dairy cows is associated with a number of behavioural changes, which can be monitored automatically. These include changes in the way the cow distributes its weight between its legs when standing, reductions in number of visits to an automated milking system, increased time spent lying down and fewer steps taken. These measures have been validated chiefly by relating them to gait scores or to the presence of hoof lesions, and each has been shown to have a reasonable degree of specificity and sensitivity in identifying lame cows, especially when different measures are combined. However, this approach shows a correlation between these measures and impaired gait but does not identify the direction of their causal relationship. This can be done by correlating changes in these behavioural measures with changes in gait that are experimentally induced by hoof trimming or treatment of lameness. A second method of validation is to examine the sensitivity of the measures to pain reduction through the use of analgesics or anaesthetics. Technological developments have provided us with a variety of tools that can be used to monitor behaviour automatically, and these have great potential to improve the detection of lame cows.

Using feeding behaviour to assess feed bunk access, competition and illness in dairy cattle

DeVries, Trevor J.[1] and Von Keyserlingk, Marina A.G.[2], [1]Dept. Animal and Poultry Science, University of Guelph, Kemptville Campus, 830 Prescott Street, Kemptville, Ontario, Canada, K0G 1J0, [2]University of British Columbia, Animal Welfare Program, 2357 Main Mall, Vancouver, Canada, V6T 1Z4

In the last decade there has been a tremendous amount of empirical research focused on feeding behavior of dairy cattle because of the obvious links between feed intake and production, as well as its use in the identification of sick animals. This has arisen in large part due to the availability of commercially available equipment that is readily available for the automated monitoring of activity at the feeder. These technologies provide ready access to vast quantities of accurate and repeatable measures including feeding frequency, duration and intake. Researchers have shown the importance of providing equal feed bunk access for dairy cattle, including the growing heifer, particularly during periods of peak feeding activity after fresh feed is delivered and when competition is highest. Work has also focused on the negative effects of competition; namely, aggressive displacements and decreased intakes. Subordinate animals (those displaced more often from the feed bunk) are most likely to return to the feed bunk during non peak feeding times when the risks is highest that there is little feed available and/or when the quality of the feed remaining is reduced. It follows that the greatest reductions in time spent feeding and intake are seen in these subordinate cows, despite their attempts to compensate by eating at a faster rate than more dominant cows. Moreover, other work has shown that those cows showing the greatest reductions in feeding time and intakes are also at greatest risk for succumbing to illness. This research has helped to identify management regimes that negatively affect feeding behaviour in dairy cattle, as well as aid in identification of cows at risk for illness. It stands to reason, therefore, that there may be utility in using automated measurement of feeding behaviour to make assessment of dairy cattle welfare in different housing and management situations.

Maternal effects on the welfare of chickens and fish

Janczak, Andrew M.[1] and Eriksen, Marit Skog[2], [1]Department of Production Animal Clinical Sciences, The Norwegian School of Veterinary Science, P.O. Box 8146 Dep. NO-0033, Oslo, Norway, [2]Department of Animal and Aquacultural Sciences, Norwegian University of Life Sciences P.O. Box 5003, 1432 Ås Norway; andrew.janczak@nvh.no

Steroid hormones produced by the mother may be incorporated into the nutritive yolk prior to egg laying or spawning and influence the development of progeny. Mildly stressing conditions may result in progeny that are better adapted to the maternal environment. More extreme stress, on the other hand, may have a range of negative consequences for welfare and productivity through adverse effects on survival, growth, adaptability, stress susceptibility, and emotionality. Hormone treatment of eggs can be used to test the effects of specific hormones on developmental processes and outcomes in chickens. We have shown that injection of corticosterone into chicken eggs prior to incubation has a number of negative consequences for behavioural development. Corticosterone-treated chicks have higher fearfulness, higher sensitivity to stress, a reduced ability to solve spatial and motor tasks, a reduced ability to compete, altered filial imprinting, lower growth, and higher fluctuating asymmetry. Psychological stress represented by unpredictable feeding of hens prior to egg laying, results in chicks that are more fearful, and have a reduced ability to compete for access to feed. We have also shown that Atlantic salmon given intraperitoneal cortisol implants have reduced fertilization. Progeny have lower initial survival rates, increased fluctuating asymmetry, impaired growth during early life, more morphological abnormalities, and altered locomotor activity in response to stress as adults. Furthermore, cortisol injection into dams prior to stripping produces offspring that have an increased percentage of unsuccessful feeding attempts, are more subordinate, receive more bites from other fish, and have altered levels of activity during acute stress. These studies indicate potential mechanisms linking the maternal environment to subsequent developmental outcomes. They also suggest the maternal environment should be taken into account in strategies aimed at improving the welfare and productivity of poultry and fish.

How perinatal and early-life experiences affect behavioural development in domestic chicks

Rodenburg, T. Bas, Animal Breeding and Genomics Centre, Wageningen University, P.O. Box 338, 6700 AH, Wageningen, the Netherlands

Perinatal and early-life experiences play a key-role in behavioural development of domestic chicks. Both the conditions under which the eggs are laid, the brooding period and the early-life environment can have profound effects. It has been shown that chicks from mothers that were stressed during egg-laying were more fearful and less competitive than chicks from control mothers, although these chicks were reared without their mother. Similar results were found if eggs were treated with corticosterone, indicating that the (stress) hormone levels in the egg influence the behaviour of the chicks. Research is currently underway to study if maternal stress is also a risk factor for the development of feather pecking in the offspring. During brooding, pre-natal light exposure can influence behavioural development: it has been found that chicks that were exposed to light for only two hours on day 19 of brooding showed higher levels of gentle feather pecking than dark-brooded chicks. This was probably due to changes in brain lateralization. The early-life environment also has strong effects on behaviour. The absence or presence of maternal care plays a central role. We have shown that maternal care during the first six weeks of life leads to changes in the peripheral serotonergic system, involved in coping with fear and stress, in adult birds. Further, it was found that maternal care leads to increased lateralization of mineralocorticoid receptors in the brains of these same birds, probably in response to reduced corticosterone levels in brooded chicks at young age. These changes are also found in behaviour: the presence of a mother results in chicks that are less fearful and show more foraging behaviour and that show less feather pecking and cannibalistic behaviour as adults. To improve welfare of domestic chickens, more attention should be given to perinatal and early-life experiences.

Prenatal life and animal welfare in the domestic pig

Rutherford, Kenneth M.D., Jarvis, Susan and Lawrence, Alistair B., Scottish Agricultural College, Work Sir Stephen Watson Building, Bush Estate, Penicuik EH 27 0PH, United Kingdom; kenny.rutherford@sac.ac.uk

That the experience of stress or under-nutrition by pregnant females affects progeny biology is widely appreciated. Such prenatal effects may guide appropriate phenotype development under natural conditions and some farm animal studies do suggest that prenatal information can produce a closer match between individuals and their postnatal environment. However, negative welfare outcomes are also common. We discuss, using the domestic pig as an example, why, given the putative adaptive basis to many prenatal effects, such negative welfare outcomes occur. Some deleterious prenatal effects may simply represent defective biological functioning or may be the consequence of trade-offs made to promote survival at the time of the challenge. Alternatively, the changes seen may have had a previous adaptive benefit that no longer functions because animals are kept in circumstances far removed from nature, i.e. prenatal effects are a form of 'evolutionary trap'. Furthermore, defensive body systems (e.g. behavioural and physiological reactivity, cognitive judgements of threat, responses to pathogens, pain) may be inherently sensitive to alteration in the direction that represents a safety first approach (i.e. lowered response thresholds, increased reactivity) with negative side-effects for animal welfare. Finally, domesticated animals may be more sensitive to prenatal effects either because their capacity to adapt to challenges has been reduced, or because maladaptive changes are no longer constrained by the requirement for survival and reproduction in the wild. In conclusion, the prenatal environment has substantial relevance for farm animal welfare. Despite the domestication process many aspects of ancestral biology remain in farm animals and may still dictate how animals respond to the captive environment. Considering both prior natural selection forces and subsequent artificial selection, provides a valuable insight into the importance of the prenatal environment for animal welfare outcomes in farm animals. Such understanding may lead to improved management of pregnant animals and their offspring.

Prevalence of keel bone deformities with an emphasis of Swiss laying hens

Käppeli, Susanna[1], Gebhardt-Henrich, Sabine G.[1], Pfulg, Andreas[1], Fröhlich, Ernst[1] and Stoffel, Michael H.[2], [1]Federal Veterinary Office, Centre for Proper Housing of Poultry and Rabbits, Zollikofen, Switzerland, [2]Division of Veterinary Anatomy, University of Berne, Switzerland

Switzerland is a leading country for alternative housing systems for laying hens. Up to now, the occurence of keel bone deformities has never been investigated. Two studies were performed to analyze the situation in Switzerland. In a prevalence study, 42 end of lay flocks of typical Swiss layers (all from percheries or aviaries) were examined at slaughter by palpation of the keel. On average, almost 55% had deformities of the keel, of which 25% were moderate or severe and can be classified as fractures. The prevalence in percheries was significantly lower than in aviaries. In an experimental study, 4,000 laying hens of a brown layer hybrid (LB) and its female parent stock were raised in an aviary system and then kept in 24 pens in a perchery under Swiss husbandry condition with perches, deep litter and nest boxes. Half of the groups received a Vitamin D3 metabolite (HyD) in the diet. Besides nutrition, perch material and genetics were examined. Every six weeks during rearing and laying period, keel bones were examined by palpation. There were practically no deformities detected during the rearing period until 18 weeks of age. With the beginning of laying, deformities increased continuously reaching 44% moderate and severe deformities on average over all groups with 62 weeks of age. HyD in the diet did not influence keel bone status. Parent stock animals had significantly fewer deformities compared with the layer hybrid (LB). Groups with metal perches had significantly more deformities than groups with plastic perches. The results of our studies show that keel bone deformities are an important problem in Swiss laying hens affecting the animals welfare. Most important factors were husbandry system, perches and genetics.

Perching behaviour and pressure load on keel bone and foot pads in laying hens

Scholz, Britta, Pickel, Thorsten and Schrader, Lars, Friedrich-Loeffler-Institut (FLI), Institute of Animal Welfare and Animal Husbandry, Dörnbergstrasse 25/27, 29223 Celle, Germany, britta.scholz@fli.bund.de

The provision of perches in housing systems for laying hens is a legal requirement from 2009 (Germany) and 2012 (EU), respectively. Although perches have a high behavioural priority for hens, their use is associated with keel bone (KB) deviations, which may possibly result from high mechanical pressure loading during extended perching activities, unstable sitting postures and perch collisions. So far, nothing is known about the pressure that acts on KB and foot pads (FP) in perching hens. In this study we analysed pressure peaks and pressure distribution between KB and FP using solid (steel, wood, pvc) and soft (rubber with air cushion) test perches of varying designs (N=14). Furthermore, the number of hens' balance movements (BM) on perches of different surface material (steel, wood, rubber) and diameter (27, 34, 45mm, N=9, 60 hens in total) were analysed during night-time observations. For pressure measurements, test perches were covered with a thin sensor film and 36 laying hens (LSL, LB) were consecutively placed on each perch. In perching hens, pressure peaks (g/cm^2) were about 4.5 times higher on KB compared to single FP. On soft perches (48mm), reduced KB pressure peaks were observed compared to four solid perch types used in practice ($356.5\pm13.2g/cm^2$ vs. 601.0 ± 10.1 to $635.3\pm11.9g/cm^2$, $P<0.001$). Furthermore, pressure peaks were lower on square compared to round ($P<0.05$) and oval, solid perches ($P<0.01$). In the following, on-farm assessments are required in order to test whether soft perches may reduce KB problems in laying hens under farming conditions. With relation to hen behaviour, BM occurred less frequently on rubber perches compared to steel and wood ($P<0.05$) and on perches of the largest diameter (45mm) compared to smaller ones ($P<0.001$), which provides evidence that an analysis of hens' perching behaviour may additionally give important information on the suitability of a perch design.

Factors affecting the movement between perches in laying hens

Haskell, Marie J.[1], Moinard, Christine[2], Sandilands, Victoria[2], Rutherford, Kenneth M.D., Statham, Poppy, Wilson, Sandra and Green, Patrick R., [1]Sustainable Livestock Systems Group, SAC, Bush Estate, Penicuik EH26 0PH, [2]Scottish Agricultural College, West Mains Road, Edinburgh, EH9 3JG

After 2012, a ban on caged systems for laying hens will mean that all laying hens in Europe will be housed in barns, aviaries or enriched cages. Many countries appear to be moving away from cages completely. While non-cage systems allow birds to show a much greater range of behaviours, they are associated with an increased incidence of bone damage. These injuries are presumed to be the result of collisions with perches or other house 'furniture'. In order to reduce the incidence of bone damage, a number of studies have explored the effect of perch arrangement, light intensity and contrast with the background on the ability of hens to move successfully between perches. Hens find wider inter-perch distances more difficult to negotiate than narrower distances (<60-75cm vs. >=80cm). A study suggested that the bird's control of the flight trajectory deteriorated as the inter-perch distance increased. Hens find upwards jumps easier than downwards jumps, and steeper angles more difficult than shallower angles (30° vs. 45° and 60°), especially for downwards jumps. At the light levels required by current European regulations (>5 lx), there are no effects of different light levels or perch contrast against the background. More difficulty in landing and flying between perches was noted when birds or other objects were present on the landing-side perch. The results suggest that taking the bird's jumping, landing and flying abilities into account when designing perch or housing systems might reduce the incidence of damaging collisions. However, there are still high levels of fractures in systems with well-designed perches. This suggests that other approaches are needed. Using breeding approaches to improve bone strength may also be required.

Are perches responsible for keel bone deformities in laying hens?

Sandilands, Victoria[1], Baker, Laurence[1], Brocklehurst, Sarah[2], Toma, Luiza[1] and Moinard, Christine[1], [1]Scottish Agricultural College, West Mains Road, Edinburgh, EH9 3JG, UK, [2]Biomathematics & Statistics Scotland, James Clerk Maxwell Building, King's Buildings, Mayfield Road, Edinburgh, EH9 3JZ, United Kingdom

Perches for hens in extensive systems are sometimes criticised as being responsible for keel bone damage. Here, we investigated three topics: a) forces that hens apply to perches when landing, b) how perch design, pen configuration, and bird genetics affect injury, and c) if keel bone damage in hens from commercial sheds differed with and without low X-frame perches. Hens that performed successful low (30 cm) and high (60 cm) downward jumps (45°) applied four to seven times their body weight force when landing, respectively. 39% of high jumps resulted in poor landings (e.g. fell off, hit keel) versus just 2% of low jumps. High and low bone strength hens (n=320) housed in 40 pens with either high (2 m) or low (60 cm) ceilings were given access to either no perch, a low solid perch, a high solid perch or a high springy perch. Half of all birds were assessed for keel bone damage by ultrasound every 8 weeks. 36% of birds with and 32% of birds without perches, and 37% of hens housed with high and 30% of hens housed with low ceilings, had keel damage by 50 weeks of age. 80 commercially-housed hens that had access to perches or not (4 blocks of 3,750 hens per treatment) were culled every 8 weeks from 17-71 weeks and assessed for keel bone damage by radiography. There was no apparent effect of perches, but the percentage of birds with damaged keels increased with bird age (from 0 to 56%). This work indicates that birds apply greater force with longer-distance downward jumps, and are more prone to injury with time, but that perches per se are not necessarily responsible for keel damage. This suggests that freedom of movement in extensive housing gives hens an opportunity to injure themselves.

Influence of access to aerial perches on keel bone injuries in laying hens on commercial free range farms

Nicholson, C.J.[1] and O'Connell, N.E.[2], [1]Agri-food and Bioscience Institute, Large Park, Hillsborough, BT26 6DR, Northern Ireland, [2]School of Biological Sciences, Queens University Belfast, Medical Biology Centre, Lisburn Road, Belfast, BT9 7BL, Northern Ireland

European legislation stipulates that laying hens are provided with 15 cm of perch space/bird. This study aimed to assess the effect of access to aerial perches on keel bone injuries, and to determine relationships between individual bird parameters and keel injury level. Treatments were applied in five commercial 8000 bird (Hyline Brown) houses. Each house (and range area) was divided in half forming 2 treatments: (1) access to aerial perches (P), and (2) no access to aerial perches (NP). Aerial perches were installed on raised slatted areas, and consisted of 6 rungs arranged at a 45° angle from the partition between the slatted and litter areas. Twenty birds/treatment were assessed at 13 points between 17 and 70 weeks of age. The birds were weighed and scores assigned for the following parameters: body condition (BCS), breast feather coverage, comb colour, body injuries, soilage of plumage and resistance to being handled. Keel bones were scored through palpation (no (0) or yes (1)). Thirty birds/treatment were randomly selected from 3 houses at 72 weeks. Body weight and comb colour were assessed before euthanasia, then breast feather coverage, BCS and girth were recorded. The keel bone was removed and injury severity scored (none (0) - severe (4)). ANOVA and REML Variance Components Analysis were used to determine treatment and age effects on keel injury parameters. Relationships between keel injury and individual parameters were assessed using Pearson correlations and Chi-Square analysis. Palpated keel injury score increased significantly across the lay cycle ($P<0.001$). There was no significant effect of aerial perches on average keel injuries measured by palpation (NP 0.335, P 0.356, SED 0.0312) or dissection (NP 1.225, P 1.472, SED 0.3440). Palpated keel score was related to breast feather coverage, combcolour, BCS, body injuries ($P<0.001$), soilage of plumage and resistance to being handled ($P<0.05$). Overall, these results suggest that aerial perches do not cause increased keel injuries in commercial free-range systems.

Conservation in cages: captive animal welfare and *ex situ* conservation

Draper, Chris[1,2], [1]Born Free Foundation, 3 Grove House, Foundry Lane, Horsham, West Sussex, RH13 5PL, UK, [2]University of Bristol, School of Biological Sciences, Woodland Road, Bristol, BS8 1UG, United Kingdom; chris@bornfree.org.uk

Ex situ conservation is an attractive, yet problematic, proposal in the face of habitat destruction and loss of biodiversity, and is a binding requirement for Parties to the Convention on Biological Diversity. It is reflected in a legal requirement for zoos in Europe to participate in conservation, although the degree of uptake varies. The conservation justifications for keeping wild animals in zoos include education, population supplementation, reintroduction and maintenance of insurance populations. However, the ability of zoos to produce and maintain behaviourally- and ecologically-competent individuals can be questioned. The impacts of captivity on animal behaviour and welfare are increasingly well-documented, and deserve special consideration under the paradigm of zoo-based conservation. Evidence from UK zoos indicates that better animal welfare performance is associated with increased participation in conservation. The causation of this relationship is unclear, but it indicates that improvements to welfare standards in zoos may improve conservation performance. However, does the finite capacity of the captive environment and its impact on animal welfare limit the contribution of zoos to conservation? Are the costs to animal welfare in captivity outweighed by the benefits to conservation? This workshop presents an opportunity to review the science of animal welfare in captivity, the legal, ethical and economic considerations, and the interplay between welfare and conservation in zoos.

A welfare construct for wild animals: can it mean anything to them?

Goddard, Pete, Macaulay Land Use Research Institute, Craigiebuckler, Aberdeen, AB15 8QH, United Kingdom; p.goddard@macaulay.ac.uk

For wild species which have had minimal or no exposure to humans, do we need to look at their welfare in a special way and develop relevant tools to evaluate their welfare? If current frameworks used for farmed species are not considered adequate or relevant for wild species, we need to develop new approaches. As we begin to impact on animals in previously undisturbed natural habitats, we begin to restrict their natural life in a range of ways; perhaps an evaluation based on such restrictions could form the starting point for a new approach. This would accord with the 'naturalness' stance which could be expanded to contain aspects of both feelings and function. Adopting this argument, we could consider what it is about any conservation action that impinges on naturalness; so what should we guard against to protect the welfare of animals involved? At least six restrictions could be imposed on the conserved species: ranging behaviour; foraging ability; breeding choice; lifespan; health; lack of disturbance. By restricting any of these and thus failing to meet the animal's needs, there is a danger that a welfare 'cost' impacts on their ability to lead a natural life. In the UK, FAWC has recently proposed that farmed species should at least be able to experience a life worth living (and ideally 'a good life') promoting positive welfare attributes rather than simply avoiding negative attributes, and that there is a need to develop a 'guardianship' role for the animals we farm. For wild species, ill-judged conservation activities could move an animal from experiencing a 'good life' to a lesser position on a welfare scale. Who should be the guardian of the welfare of animals in the wild? We hope to explore further the utility of this approach in the workshop.

Reintroduction success: should this be planned at the individual or species level?

Bremner-Harrison, Samantha, Nottingham Trent University, Brackenhurst Campus, School of Animal, Rural & Environmental Sciences, Southwell, NG25 0QF, United Kingdom; samantha.bremnerharrison@ntu.ac.uk

Reintroduction and translocation are common tools used in the conservation of threatened or endangered species. However, for every successful release programme there are many more that do not succeed for various reasons. Even in those programmes that are deemed a success mortality levels are generally high, particularly during the early stages of the programme or during the initial post-release period. Programmes that utilise wild-caught individuals animals are generally considered to have greater chances of survival however even here there are significant levels of mortality. Successful release programmes undoubtedly result in the recovery of species that may have otherwise become locally or globally extinct, and in the case of keystone species may have ultimately benefited a range of dependent species. However, these programmes all have some level of mortality associated with them. The inclusion of personality assessments for release candidates may limit the loss of individuals, however, implementation of this additional step to a release strategy may add to the overall cost and length of the programme. Animal welfare scientists and conservation planners may view this added cost from opposing viewpoints. Conservation managers are generally working under the constraints of limited time and financial budgets and may not view assessing the suitability of individuals for release as a step worthy of inclusion. Animal welfare scientists may argue that a measure that increases the likelihood of survival of each released individual is one that is both necessary and morally apt. This talk will discuss the potential advantages and disadvantages of including personality research in reintroduction/translocation planning, both at the individual animal and programme level.

The welfare of extensively managed animals: an interface between captive animals and wildlife?

Dwyer, Cathy, Sustainable Livestock Systems, SAC, King's Buildings, West Mains Road, Edinburgh, United Kingdom; cathy.dwyer@sac.ac.uk

Extensively managed livestock, particularly those managed very extensively such as hill sheep or range cattle, although still the responsibility of the producer, can lead a semi-wild existence for much of their lives. The environment in which they live may be highly managed, but the animals themselves experience little direct human contact or interference. In this respect, these animals are perhaps on the margins between a livestock species and wildlife, and issues relating to their welfare are also relevant to wildlife. For these animals, the opportunities to express natural behaviour can be much greater than for more intensively managed species. However, other aspects of good welfare, such as provision of food or protection from disease, are considerably more variable and can be severely compromised. In assessing the welfare of extensively managed species we need to understand how important the expression of evolved behaviour patterns is for good welfare when set against the potential for other welfare challenges. Our actions in selecting for particular characteristics in livestock may affect this balance between positive and negative aspects of welfare and control over reproductive behaviour, for example, means that livestock species do not have complete freedom of behavioural expression. These effects may be paralleled in wildlife species where, for example, hunting and the selective removal of particular animals can alter the genetic makeup of wildlife populations. Assessing the welfare of extensively managed livestock is a complex issue as the heterogeneity of extensive environments, seasonality of welfare challenges and infrequent handling of these animals all hamper the application of welfare assessment protocols developed under more intensive conditions. Further, the extensively managed animal faces welfare challenges, such as predation, to which intensively managed animal are not exposed. The development of welfare assessment protocols in extensively managed species will have direct relevance to assessing welfare in wildlife.

Wild animals as pets: animal welfare and conservation implications

Koene, Paul, Department of Animal Sciences, Wageningen University, P.O. Box 338, 6700 AH Wageningen, the Netherlands; paul.koene@wur.nl

Many animals are kept as pets; domesticated or wild ones. But are all animals suitable as companion animals? Ten years ago a framework for assessing this suitability was published by Schuppli and Fraser (2000). Part of that framework was – limited – attention to the natural behaviour of companion animals and the consequences when performing natural behaviour (for example, flying in parrots is often not possible). The trade in (wild) animals as potential companion animals is not diminished. Many people continue to keep exotic species, directly conflicting individual animal welfare in their homes and species conservation in the wild. In order to estimate the suitability of an exotic species as companion animals, information on natural behaviour is essential, as is information about the behaviour, welfare and care needed as pet animals. However, often the animal's life under natural conditions is not known. Furthermore, information about the animals as pets is not known, either because information is not published or not known, or because the animals are not allowed to be kept. In such cases, some method of predicting the welfare risks of an animal as a potential companion animal should be available. In a project to enhance the transparency and objectivity to detect the suitability of exotic species as pet animals in an early stage, a new framework is designed in which animal welfare and natural behaviour, and especially behavioural needs of animals, play a crucial role. The method includes the assessment of behavioural needs of an animal species (in 8 functional behavioural classes) and – based on the needs – the welfare risks in the domestic environment. In the last stage, this information is combined with other legal and risk factors to provide the final assessment of the potential of an animal species as a companion animal.

Authors index